SOCIAL STABILITY: THE CHALLENGE OF TECHNOLOGY DEVELOPMENT 2001
(SWIIS '01)

A Proceedings volume from the 8th IFAC Conference,
Vienna, Austria, 27 - 29 September 2001

Edited by

P. KOPACEK
Institute for Handling Devices and Robotics,
Vienna University of Technology,
Vienna, Austria

Published for the

INTERNATIONAL FEDERATION OF AUTOMATIC CONTROL

by

PERGAMON
An Imprint of Elsevier Science

ELSEVIER SCIENCE Ltd
The Boulevard, Langford Lane
Kidlington, Oxford OX5 1GB,UK

Elsevier Science Internet Homepage
http://www.elsevier.com

Consult the Elsevier Homepage for full catalogue information on all books, journals and electronic products and services.

IFAC Publications Internet Homepage
http://www.elsevier.com/locate/ifac

Consult the IFAC Publications Homepage for full details on the preparation of IFAC meeting papers, published/forthcoming IFAC books, and information about the IFAC Journals and affiliated journals.

First edition 2002

Library of Congress Cataloging in Publication Data

A catalogue record for this book is available from the Library of Congress

British Library Cataloguing in Publication Data

A catalogue record for this book is available from the British Library

ISBN 0-08-043961 6
ISSN 1474-6670

Transferred to digital printing 2005

These proceedings were reproduced from manuscripts supplied by the authors, therefore the reproduction is not completely uniform but neither the format nor the language have been changed in the interests of rapid publication. Whilst every effort is made by the publishers to see that no inaccurate or misleading data, opinion or statement appears in this publication, they wish to make it clear that the data and opinions appearing in the articles herein are the sole responsibility of the contributor concerned. Accordingly, the publisher, editors and their respective employers, officers and agents accept no responsibility or liability whatsoever for the onsequences of any such inaccurate or misleading data, opinion or statement.

To Contact the Publisher

Elsevier Science welcomes enquiries concerning publishing proposals: books, journal special issues, conference proceedings, etc. All formats and media can be considered. Should you have a publishing proposal you wish to discuss, please contact, without obligation, the publisher responsible for Elsevier's industrial and control engineering publishing programme:

Dr Martin Ruck
Publishing Editor
Elsevier Science Ltd
The Boulevard, Langford Lane
Kidlington, Oxford
OX5 1GB, UK

Phone:	+44 1865 843230
Fax:	+44 1865 843920
E.mail:	m.ruck@elsevier.co.uk

General enquiries, including placing orders, should be directed to Elsevier's Regional Sales Offices – please access the Elsevier homepage for full contact details (homepage details at the top of this page).

8TH IFAC CONFERENCE ON SOCIAL STABILITY: THE CHALLENGE OF TECHNOLOGY DEVELOPMENT 2001

Sponsored by
International Federation of Automatic Control - IFAC

IFAC – TC on Supplemental Ways for Improving International Stability (SMW)

Organized by
Institute for Handling Devices and Robotics - IHRT, Vienna University of Technology

Co-sponsored by
IFAC Technical Committees on

-   Developing Countries (GEA)
-   Social Impacts of Automation (GES)
-   Computation in Economic, Financial and Engineering-Economic Systems (SME)
-   Business and Management Techniques (SMB)

International Federation for Information Processing (IFIP)

International Federation of Operational Research Societies (IFORS)

International Measurement Confederation (IMEKO)

Vienna University of Technology (VUT)

CA - Creditanstalt Bankverein Austria (CA-BV)

International Programme Committee (IPC)
Kile F. (USA) (Chairman)

Antonini, C. (ZA)	Kopacek, P. (A)
Cernetic, J. (SLO)	Lu, Y.Z. (PRC)
Cioffi-Revilla, C. (USA)	Mansour, M. (CH)
Dimirovski, G. (MAC)	Martensson, L. (S)
Dinibütün, A.T. (TR)	Richardson, J. (F)
Dumitrache, I. (RO)	Scheffran, J. (D)
Groumpos, P.P. (GR)	Thoma, M. (D)
Hersh, M. (UK)	Vamos, T. (H)
Holubiec, S. (PL)	

National Organizing Committee (NOC)
Kopacek, P. (A) (Chairman)

Han, M.W. (A)

Nemetz, I. (A)

Zebedin, H. (A)

FOREWORD

First ideas on Social Stability in IFAC came up in the early 80s stimulated by Hal Chestnut, the first president of IFAC. Hal Chestnut passed away at the end of August 2001 and therefore, this 8th IFAC Conference on "Social Stability: The Challenge of Technology Development (SWIIS´01) is dedicated to him. Hal Chestnut founded in IFAC the Technical Committee on this subject and organized the first few IFAC Workshops on this topic: Laxenburg, Austria (1983), Cleveland, Ohio, USA (1986), Budapest, Hungary (1989) and Toronto, Ontario, Canada (1992). The last two regular conferences took place in Sinaia, Romania (1998) and in Ohrid, Macedonia (2000). The last one was more or less an intermediate workshop.

The technological development has caused profound changes to social stability. Regions that had stable populations for centuries have experienced enormous population growth leading to the emergence of sometimes-unmanageable megaplex cities as well as bringing about macroscopic environmental change. Therefore, the scope of this IFAC SWIIS conference is to offer insights into mitigating unwanted side-effects of rapid development and to share methodologies for appropriate ways of managing the introduction of technologies which will alter social stability.

The papers included in this Proceedings volume cover a very broad field of interest for those subjects like social aspects of technology transfer, managing the introduction of technological change, ethical aspects, technology and environmental stability and anticipating secondary and tertiary effects of technological development.

On behalf of the International Program Committee (IPC) and the National Organizing Committee (NOC), we would like to thank all the participants for contributing to this event and hope the conference stimulated intensive and successful discussions.

Vienna, September 2001

Peter Kopacek
Editor

This Conference is dedicated to

Harold Chestnut,

Founder of SWIIS

CONTENTS

VARIOUS

IFAC

Publications
www.elsevier.com/locate/ifac

WHISTLEBLOWERS - HEROES OR TRAITORS?: INDIVIDUAL AND COLLECTIVE RESPONSIBILITY FOR ETHICAL BEHAVIOUR

M.A. Hersh

Centre of Systems and Control and
Department of Electronics and Electrical Engineering,
University of Glasgow, Glasgow G12 8LT, Scotland.
Tel: +44 141 330 4906. Fax: +44 141 330 6004. Email: m.hersh@elec.gla.ac.uk

Abstract: This paper reviews the literature on whistleblowing in the context of the ethical issues and conflicts of loyalties it raises. Factors which effect the likelihood of whistleblowing, such as individual and organisational characteristics and the severity of the incident, are discussed. Organisational responses, including retaliation, and the effectiveness of whistleblowing are considered, as well as the state of legal protection in the US and UK. The particular issues raised by whistleblowing in science and research are considered and the similarities and differences in the treatment of whistleblowers in the former Soviet Union and the US examined. *Copyright ©2001 IFAC*

Keywords: Whistleblowing, ethics, retaliation, legislation, responsibility

1. INTRODUCTION

Whistleblowing involves the deliberate disclosure of information about non-trivial activities which are believed to be dangerous, illegal, unethical, discriminatory or to otherwise involve wrongdoing, generally by current or former organisation members. The term 'whistleblowing' was first used in the 1963 publicity about Otto Otopeka (Petersen et al, 1986; Vinten, 1994a), who had given classified documents about security risks in the new US administration to the chief counsel of the Senate Subcommittee on Internal Security. The term is apparently derived from English policemen blowing their whistles to alert the public and other police to criminal acts (Strader, 1993). Whistleblowing has been discussed in official reports by the OECD (2000) and in Australia (EARC, 1990), Canada (Ontario, 1986) and the US (Leahy, 1978).

There are several review papers of whistleblowing, including a review of the early literature and resource materials by Bowman (1983) and a more recent review by Miethe et al (1994). Science and Engineering Ethics recently had a special issue on whistleblowing (Sci, 1998), which considered issues such as the psychology of whistleblowing, the scientific community's responses, personal experiences of whistleblowing and advice to whistleblowers for maintaining a career afterwards.

While there is some consensus on many features of whistleblowing, there is no universally agreed definition (Jensen, 1987; Judd, 1999; Bernstein et al, 1996; Near et al, 1985). Most of the definitions agree that whistleblowing involves the reporting of questionable morality and/or wrongdoing which is not confined to illegality. Disagreement relates to actor and recipient attributes, such as membership of the organisation being criticised, the circumstances of the disclosure and motive and whether disclosures must be external and unauthorised or can be internal and/or permitted to count as whistleblowing.

Most whistleblowers act on their own. De Maria (1992) suggests that this lone voice aspect often puts them in a particular conservative political context and can allow them to be recruited back into the system through internal disclosure. Bok (1981) concurs in this view that open door policies to encourage internal disclosure can turn into traps if the abuse is planned by those in charge. De Maria (1992) proposes that governments should encourage collectivised workplace dissent or whistleblowing as a class action in addition to protection for individual whistleblowers. However it is unlikely this approach will be adopted by governments or organisations trying to coopt whistleblowers and limit the effects of their disclosures to correcting specific abuses.

Attitudes to whistleblowers vary, as indicated by the terms used to describe them, such as conscientious objector, ethical resister, informer and licensed spy (Vinten, 1994). On the one hand there is a belief that whistleblowing is an ethical or even praiseworthy act, which is required to expose abuses of all kinds and avoid moral complicity in them. On the other hand whistleblowers may be seen as informers who betray colleagues and the organisations they work for. A particularly negative view of whistleblowing is expressed by Drucker (1981), who equates it with informing and gives examples of violent tyrannies that encouraged informers. However, the tone of the majority of articles is supportive of whistleblowers.

Whistleblowing is sometimes seen as a US phenomenon. However there is also a body of literature on whistleblowing in the UK and a smaller body of literature on, for instance, Australia (Caiden et al, 1994; De Maria et al, 1996; De Maria, 1999; Tucker, 1995), Hong Kong (Chua, 1998; Clark, 1994; Lui, 1988) and Russia (Von Hippel, 1993).

2. ETHICAL TENSION POINTS

Whistleblowing involves conflicts of loyalties and ethical tension points which have been divided (Jensen, 1987) into procedural and substantive. Procedural points include the gravity of the problem, information handling issues, motivation, anonymous versus open whistleblowing, the appropriate audience, and whether the whistleblowing act will be worth the costs to the whistleblower and accused in terms of time, money, effort and mental involvement.

Jensen (1987) and Judd (1999) both consider the main ethical dilemma in whistleblowing to be the conflicts involved in balancing values, multiple loyalties and obligations to the organisation, the general public, professional associations, family and friends and oneself. Judd (1999) also considers the relationship of whistleblowing to and differences from informing and dissent. Jensen (1987) suggests that whistleblowers challenge the assumption that what is good for the organisation is good for the

wider public. Bormann (1975) highlights the need for choices about loyalties when there are conflicts between group and society norms and ends. Jensen (1987) recognises that attitudes to group loyalty vary in different cultures. Devine (1995) suggests that whistleblowers are at the intersection of valid but conflicting fundamental values, including conflicts between the right to privacy and the public's right to know. Bernstein et al (1996) consider the violation of professional standards to be the most common cause of whistleblowing. This tension between the need to prevent abuses and to preserve trust is an important tension point in whistleblowing and a major source of ambiguity about it. Laframboise (1991) has suggested that the abhorrence of an act to peer group values rather than its illegality or impropriety determines whether whistleblowing is perceived as justified. However in some cases peer group values may accept behaviour which is, for instance, damaging to individuals, minority groups and/or the environment so that whistleblowing may be required to challenge it. There may also be tensions between consequentialist (based on likely consequences) and deontological (based on the intrinsic morality of an action) ethics, for instance whether minor wrongdoing can be justified to prevent severe consequences.

A number of authors stress the value of loyalty to the employing organisation and conflicts between this loyalty and duties to the wider society. Larmer (1992) discusses the type of loyalty owed by employees and suggests a definition of loyalty based on acting in accordance with what one considers to be the person's best interests, which could on occasion involve acting against their wishes. However this type of argument both sidesteps the emotional impacts of conflicts of loyalties (however defined) and conflicts between confidentiality and the duty of public disclosure. There seems to have been little research to test the assumption of loyalty to the organisation and whether its existence or degree is dependent on other factors, such as gender or belonging to a minority group. It is quite possible that women and minorities feel less loyalty or consider themselves as outsiders in many organisations and this may be one of the reasons for the lower reported whistleblowing amongst women.

Some managers consider that the duty of confidentiality to the organisation should override ethical or other concerns, and that employees with ethical concerns should resign (Winfield, 1994). However this is a high price to pay for behaving ethically and may be difficult or impossible for individuals with heavy financial commitments or dependants. Unless accompanied by whistleblowing, resignation is unlikely to effect the activity about which there is concern and may not even prevent retaliation. It is also open to question whether resignation without whistleblowing resolves the question of moral complicity.

Different types of employment often have their own specific conflicts and ethical issues. For instance in the medical and related professions there are possible ethical conflicts between publicising information about, for instance, inadequate, inappropriate, incompetent or dangerous treatments and protecting patient confidentiality, since reports are likely to be more effective if they include details of specific patients. Informing on colleagues also raises particular ethical dilemmas. In addition professional codes and practice may promote excessive loyalty to members of the profession and protection of them which can damage patients (Baab et al 1994). There are also issues of how genuine differences of scientific and medical opinion should be presented to members of the public and the extent to which medical professionals should be expected to sacrifice their own interests for their patients (Edwards 1996).

There are a number of well known examples, such as Bhopal and Chernobyl, where whistleblowing could have averted disasters, with the consequent loss of life and environmental damage. However, the power and wealth of big business in particular can make it difficult to take a stand. In addition workers in private industry may have less job protection than those in the public sector. Environmental, social and health impacts of industrial activity frequently extend beyond the borders of one state, but legal protection may be limited to whistleblowing about concerns within the boundaries (Sternberg, 1996).

3. THEORIES OF WHISTLEBLOWING

Most research on whistleblowing has been descriptive or considered some, but not all the characteristics of whistleblowing. Miethe et al (1994) consider that theories should incorporate personal characteristics of the observer, the situational context and the organisational structure, whereas the author considers that the type and severity of the activity of concern should be considered as an additional separate category. A number of social psychological theories such as motivation, resource dependency and reinforcement theory have been applied to whistleblowing, but their validity has not been empirically tested. Miethe et al (1994) suggest explanations of whistleblowing behaviour which use social learning, social bond and rational choice theory and that a crucial factor is whether whistleblowing is considered deviant or conventional behaviour. Anderson et al (1980) discuss the organisational conditions that give rise to concern or disagreement with organisational practices and try to put whistleblowing in the larger economic, political and professional context.

There have been several investigations of the characteristics of whistleblowers and/or workplace structures which lead to whistleblowing. Unfortunately few researchers have considered the interaction between individual and organisational characteristics, type of wrongdoing and situational factors and most have investigated only individual characteristics or organisational structure. Hypotheses tend to be one rather than multi-dimensional and seem to ignore the fact that there can be different groups or types of individuals who become whistleblowers, for instance both loners/outsiders and conventional organisational identified individuals. The focus on whistleblower characteristics also makes the assumption, which has not been empirically tested, that there is a particular type of person who becomes a whistleblower, though there are more likely to be several whistleblower types. Though there may be personal characteristics that are strongly correlated with whistleblowing, it is probable that many or even most people could act as whistleblowers in the right circumstances. There do not seem to have been any attempts to categorise the range of situations that can result in whistleblowing in terms of consequences, ethical issues, conflicts of loyalties and other factors.

4. ORGANISATIONAL CHARACTERISTICS

There are generally advantages to organisations in encouraging internal whistleblowing and correcting abuses without the need for external disclosure Barnett et al (1993). A number of authors have carried out surveys to investigate organisational characteristics which influence whistleblowing behaviour and the ethical climate of the organisation. The surveys vary in size, percentage response rate and sector of the economy, but the results seem to show some consistency. Low response rates to surveys on ethical policies and performance, for instance (Winfield, 1994), may indicate that ethics are not a significant concern for many companies, though this is probably changing. Measures which could encourage ethical behaviour in organisations include a code of ethics, a whistleblowing system and ethics focused decision making (Lindsay et al, 1996).

Suggestions for reducing external disclosures include the development of internal disclosure policies and procedures, with appropriate communication channels, formal investigative procedures and guarantees of protection for good faith disclosures (Barnett et al, 1993); clear guidelines on acceptable and unacceptable behaviour, monitoring procedures and clear procedures for raising concerns, anonymously if preferred, at the highest level (Winfield, 1994); and cultural change with strong endorsement of ethical conduct and discussion of ethical issues by management, punishment for violations, and effective protection for whistleblowers (Benson et al 1998). However a survey of first level managers (Keenan, 1995) found that over half were uncertain of the protection offered by their firms to whistleblowers and a third were not confident that whistleblowers would not experience reprisals. Fear of retaliation and knowing where to report were the main negative and positive factors in whistleblowing behaviour.

Although the evidence is not totally conclusive, surveys have found that larger companies and unionised companies perceive higher levels of external disclosure (Barnett, 1992) and that companies with disclosure policies have higher rates of both internal and external disclosures than those without (Barnett, 1993), possibly because employers adopt policies after negative experiences with whistleblowing. Firms with a codes of ethics, formal participation policies and committees for listening to employees seemed more aware of the importance of listening to employee concerns (Winfield, 1994). The following organisational characteristics generally encourage external whistleblowing: indirect and complex lines of communication and authority and discouragement or suppression or expressions of doubt or technical and other dissent (Perrucci et al, 1980; Westin, 1981; Elliston et al, 1985), lack of knowledge of internal communication channels (Miceli et al 1984) and complicated hierarchies (King, 1999). However even relatively open organisations can experience communication blockages (Anderson et al, 1980) which could lead to external whistleblowing. Survey also show (Callahan et al 1992) that most employees recognise a hierarchy of proper whistleblowing outlets, internal followed by law enforcement agencies, with news media last and that the majority of employees, including managerial and supervisory employees, support legal protection of whistleblowers, but more strongly for illegal than unethical activities.

Brock (1999) presents the Hanford Joint Council for Resolving Employee Concerns as a positive example of alternative dispute resolution principles (ADR). He highlights the fact that previously resolution of cases through litigation or settlement was expensive, but only rarely resolved the safety issues involved. The system is based on an ADR and an eight member Council with a neutral chair, two members from public interest community groups, two from the main Hanford contractors, a former whistleblower and two neutral leaders from the business, academic or labour communities. An informal network allows cases to be referred before they become polarised. Any existing procedures are put on hold with the support of the courts and other agencies. Brock considers the main factors in the Council's success to be its membership composition and tools, including the right level of authority and perspective and an agreement to implement consensus decisions.

Benson et al (1998) present the Sundstrand Corporation in Illinois as a viable model for cultural change. A penalty settlement of $227.3 million, resulting from charges of ethical violations in the mid-80s, led to reassessment with a commitment of resources and the setting up of the new post of

Corporate Director, Business Conduct and Ethics. This director became the port of call for questions about ethical conduct and reports of unethical conduct. A code of business conduct and an ethics booklet were produced and supported by a well publicised training programme. A hot line was set up with all calls receiving a response within 24 hours. The fact that both these examples of positive approaches occurred in industries (nuclear and 'defence'), the very nature of which raises ethical questions, illustrates the fact that whistleblowing generally challenges wrongdoing within the system rather than the nature of the system itself.

5. WHISTLEBLOWER CHARACTERISTICS

Although there are a number of surveys and case studies of individual whistleblowers, the lack of an obvious sampling frame complicates systematic studies. The problems in identifying whistleblowers mean that surveys are either dependent on self-selected groups (Jos et al, 1989) or responses to scenarios. Thus surveys may be unrepresentative or only representative of certain subgroups of whistleblowers and responses to scenarios may differ from those in real situations. Survey results are often limited by the small survey size and restriction to one sector of the economy and there have been few if any international comparisons or comparisons across different sectors of the economy. The survey design may also be limited, for instance, by restricting the available responses to no action or external whistleblowing (excluding internal whistleblowing) or assuming a one stage process, rather than allowing both one and multi-stage whistleblowing processes.

Empirical research on characteristics which distinguish whistleblowers has focused on demographic factors such as age, gender, social class and psychological factors such as self efficacy, locus of control and moral development (Miethe et al, 1994). Whistleblowers are often characterised as principled individuals with strong moral convictions, high levels of moral development, universal standards of justice and self-efficacy and high levels of internal control. Evidence of demographic factors is mixed (Miethe et al, 1994). Some groups of whistleblowers have been found to have a distinctive approach to moral issues and decision making and a commitment to particular values, which allowed them to act against strong organisational and situational pressures (Jos et al, 1989). Some research indicates that women are less likely to be whistleblowers than men (Miceli et al, 1992) and it has been suggested that this is due to them being more likely to conform to a majority opinion (Costanzo et al, 1966). However women frequently have less secure and lower status positions than men and therefore are likely to be more vulnerable to retaliation. Since there are more men in senior positions women's concerns may be taken less seriously. This combination of being less likely to obtain a positive response and more likely to experience retaliation could explain many of the observed differences in whistleblowing behaviour. This hypothesis is supported by Miceli et al (1992) who suggest that the gender of complaint recipient and whistleblowers may be correlated and women may be less knowledgeable about reporting channels.

Elliston et al (1985) have found that a strong sense of professional responsibility and/or commitment to the organisation's formal goals or successful completion of the project and identification with the organisation are likely to lead to whistleblowing, unless (Hacker, 1978) this identification leads to executive ambitions which will dampen whistleblowing. On the other hand loners are more likely to withstand group pressures for conformity that may prevent whistleblowing (Greenberger et al, 1987). It has been suggested (De Maria et al, 1997) that whistleblowers start off as system sympathetic people and only change their views when they experience reprisals, agreeing with other evidence that they are devoted to their work and organisations and successful until asked to violate their own ethical standards (Glazer et al, 1989). Surveys have found that external whistleblowers tend to have less tenure and greater evidence of wrongdoing, are more effective in achieving change, but experience more extensive retaliation, than internal whistleblowers (Dworkin et al, 1998a). Contrary to expectations low submissiveness to authority and self-righteousness have not been found to be predictors of whistleblowing (McCutcheon, 2000). One survey found that the overwhelming majority initially used internal outlets, and the majority then continued to external and/or public disclosure, probably due to frustration with the speed and/or quality of the internal processes and sometimes to clear their names and protect their careers (De Maria et al, 1997).

An investigation of whistleblowing behaviour in the context of corruption (Gorta et al 1995) found that public sector employees do not share a common understanding of corrupt behaviour, though there was correlation between considering behaviour corrupt and harmful, unjustified and undesirable. This may indicate problems with researchers imposing their own value categories on subjects, who may interpret them differently. They also found that some individuals would take action about behaviour they did not consider corrupt and/or not take action about behaviour they considered corrupt. The main factors that would influence decisions to take no action were beliefs that the behaviour was justified in the circumstances, reporting would not lead to action and that the scenario was not corrupt, as well as concern about retaliation and lack of supervisory status.

6. ORGANISATIONAL RESPONSES

Many surveys (De Maria et al, 1996, 1997; Glazer et al, 1989; Jos et al, 1989; Lennane, 1993; Soeken et al, 1987) and other accounts of whistleblower experience show that most whistleblowers experience retaliation, sometimes of a very severe kind. However problems with small samples, often of specific groups of whistleblowers, indicate that conclusions may be limited to the particular type of whistleblower. The severity of the potential risk is highlighted by the case of Stanley Adams, a former executive of the Swiss pharmaceutical firm Hoffman La Roche, who was imprisoned for exposing the firm's illegal price fixing methods to the European Commission in 1973 and whose wife committed suicide (Vinten, 1994).

There are indications of increasing retaliation against both relatively powerless employees and powerful but influential employees (Pamerlee et al, 1982), possibly due to likely degree of damage (Near et al, 1986) and reduced retaliation with perceived top management support and the merits of the case, but not perceived effectiveness (Near et al, 1983). A distinction (De Maria et al, 1996) has been made between official retaliation, in which punishment is covered up by policy and procedures to avoid charges of victimisation, and unofficial reprisals. 71% of one survey sample experienced official reprisals and 94% unofficial ones (De Maria et al, 1996), with multiple acts of reprisals in most cases. Formal reprimand was the most common official reprisal, followed by punitive transfer and compulsory psychiatric or other referrals. Dismissal occurred in 8% of cases. Workplace ostracism was the most common form of unofficial reprisal, followed by personal attacks and increased scrutiny. Another study found that managers who sacked internal whistleblowers generally acted very quickly, possibly after trying to silence or discredit them, but waited longer before firing external whistleblowers and tried to use nullification or isolation to silence them first, possibly as they have more evidence of wrongdoing (Dworkin et al, 1998a).

Another survey (Soeken et al, 1987) of 87 US whistleblowers from the civil service and private industry found that only one of them had not experience retaliation and harassment from peers and/or superiors. The majority in private industry and half in the civil service lost their jobs; 17% lost their homes; 8% filed for bankruptcy; 15% were divorced and 10% attempted suicide. Another study (De Maria et al, 1996; 1997;) found that 74 internal disclosures generated 104 (49%) negative, 41 (19%) obstructive and 69 (32%) procedurally correct organisational responses. The most common negative response was inaction. In 14 cases wrongdoing was substantiated but covered up by a superior and in 13 cases buck passing occurred.

7. WHISTLEBLOWING IN SCIENCE AND RESEARCH

Whistleblowing in science and research raises a number of issues, many but not all of which are shared by other areas in which whistleblowing occurs. A particular difficulty is the fact that reporting will often be of colleagues rather than management. Although trade union type solidarity is not particularly noticeable amongst researchers, there is still often a feeling of collegiality. Some of the consequences of wrongdoing in science may have wider implications, for instance in terms of implying a drug is safe and/or effective when it is not, whereas others may be unethical, but not have wider consequences. There may also be genuine disputes between scientists which do not involve misconduct, (Durso, 1996). However a particular problem, which is rarely highlighted in the context of whistleblowing, is the issue of gatekeeping with, for instance, women and ethnic minorities having less access to grants and publication in prestigious journals and difficulties being encountered in publishing theories which challenge accepted orthodoxies. There are also issues of indicating that results are controversial, particularly in areas such as genetically modified organisms which could have significant implications for the general public.

Professionals such as scientists are often expected to regulate themselves and this is often dependent on reporting. However publicised misconduct cases often reveal (Wenger et al, 1999) that reporting has not occurred. Whistleblowers may face sanctions from peers and/or the organisation. Sanctions from peers can often be professionally damaging as well as personally painful (Barnett et al, 1996). Cases of false or malicious accusation are considered to be rare (Edsall, 1995), but genuine, but unfounded accusations do occur. Unjustified whistleblowing can have damaging effects, including on individuals who are exonerated of misconduct (Klotz, 1998). Concern has also been expressed that whistleblowing could affect the public standing of science.

A focus group study (Wenger et al, 1997) of the normative ethical views of scientists and institutional representatives found that the scientists perceived strong agreement in the scientific community about norms relating to honesty, integrity and working towards the common good, but believed there would be disagreement about what is the common good. The institutional representatives strongly held that scientists have a duty to report scientific misconduct, whereas scientists feared being whistleblowers due to lack of support by the institution or other scientists. Both groups considered that attitudes to norms of scientific behaviour vary considerably across cultures and disciplines. The extreme competition for funding, which creates pressures not to communicate, not to be thorough and to publish too early was considered to be one of the main factors contributing to violations. This indicates that improving funding could be the best way to reduce misconduct.

There have been a number of studies of both actual whistleblowing and scenario behaviour, for instance (Barnett et al, 1996; Braxton et al, 1996; Lubalin et al, 1999), but many of the findings are limited by sample restriction to one field of science. Wenger et al (1999) found that the overwhelming majority of their 606 scientist respondents would report unethical behaviour, but in a third of cases to the researcher and/or colleagues. This reporting would be educational or warning and unlikely to lead to a disciplinary response. The authors question whether this behaviour is self-regulation or a cover up. However it may be an attempt to reconcile conflicting loyalties and avoid sanctions while behaving ethically. An informal educative approach may be appropriate in all but the most serious cases of misconduct. Experiences of retaliation are similar to those of other types of whistleblowers. In a study of scientist whistleblowers and accused but exonerated scientists, nearly a quarter of the scientists were fired or did not have contracts their renewed. 8% of the exonerated were fired or did not have their contracts renewed; significant minorities were denied promotion or salary increases or lost research resources or opportunities; and the mental health of the majority suffered (Lubalin et al, 1999).

Concerns about drug companies trying to withhold or manipulate results or block publication of unfavourable studies have led to a number of prominent medical journals reserving the right to refuse to publish drug company sponsored studies unless the researchers involved are guaranteed scientific independence (Okie, 2001). Researchers at the University of California recently defied a corporate sponsor by publishing a study concluding that a particular HIV therapy product does not benefit patients already receiving standard treatments. The firm is seeking $7-10 million damages from the university for harming its business. A University of Toronto physician lost a research contract with a Canadian drug company after publishing an article about a serious side effect of one of their drugs. The company claims that she failed to follow the protocol specifying how the study should be carried out.

In the US the False Claims Act allows whistleblowers to sue universities and scientists on behalf of the government for recovery of up to three times the amount of any fraudulent claims, with a percentage going to the successful claimant. In two cases of plagiarism and data falsification whistleblowers received substantial percentages of the combined penalties of more than $3 million (Hoke, 1995). The possibility of large cash awards penalising universities for scientific misconduct by researchers may lead to behavioural changes at the administrative levels, but does not challenge the underlying problems of access to funding and power structures within research. In another case a biomedical engineer filed under the False Claims Act after dissatisfaction with the results of university and funding body inquiries. He had been accused of unethical practices: he had failed to cite the paper of a colleague reporting the effectiveness of a drug, manufactured by one of his funders, whereas the whistleblower found no efficacy using the same data (Hoke, 1995). This case illustrates how easily attention can be sidetracked from substantive ethical concerns to more minor issues of professional etiquette.

Goldbeck-Wood (1997) discusses calls for a UK organisation similar to the US Office of Research Integrity or the Danish national committee for scientific dishonesty to investigate claims of scientific misconduct and impose sanctions. However Parrish (1997) questions the role of the Office of Research Integrity after the overturning and withdrawal of several of its findings of scientific misconduct. Particular problems are due to the fact that members of the appeals board are lawyers without a scientific background or understanding of scientific culture. For instance judges on the appeals board considered it acceptable to change numbers and results for the purposes of publication, as long as this did not affect the study outcome, whereas this would be unacceptable to scientists (CRI, 1994).

McKnight (1998) discusses the use of ethics policies in promoting ethical conduct in scientific societies and the way different codes treat the responsibility to expose misconduct. She suggests that a scientific society can further 'peer policing' by establishing an ethics committee and encouraging or assisting whistleblowers to report to this committee. Other proposals include more explicit guidelines for reporting and punishment (Bayles, 1981; AMA, 1997) and a witness bill of rights designed to protect the scientific community and ensure the free flow of information (Devine, 1995). Gunsalus (1998b) suggests that university administrators develop a non-defensive internal culture which does not consider a problem as an indictment of the whole institution. He provides a number of specific guidelines, such as setting boundaries with regards to time and topics, to avoid confusing personal and professional roles, and making clear any responsibilities to act on or report information received, which could interfere with maintaining confidentiality. He also suggests hearing at least two sides, not taking problems personally, stating clearly what action will be taken over what time frame and recognising when problems require more formal procedures. However these proposals all target abuses within the system, rather than wider issues such as the ethics of military research or vivisection.

A number of case studies have been developed, for instance for use in courses on research ethics. One study (Sims, 2001) presents a three part scenario with discussion questions and commentary: a doctoral student, Sherry, discovers that her supervisor has submitted a paper based on her as yet incomplete research. Although a number of real-life ethical dilemmas are presented, the focus is solely on the student and not on the supervisor's responsibilities for monitoring and teaching ethical research practices (Johnson, 2001). The context of the threat to laboratory funding and relative importance of data points and the jobs of Sherry's coworkers could have lead to an examination of the wider context of the ethical temptations posed by inadequate funding systems and the ethics of committing an unethical act in order to avoid much more serious consequences.

8. CODES FOR WHISTLEBLOWERS

A number of individuals and organisations have derived codes or sets of rules for whistleblowers to be used in decision making about whether to blow the whistle as well as throughout the process to increase the likelihood of a successful outcome and retaining a career. For instance Gunsalus (1998a) suggests that prospective whistleblowers should consider alternative explanations and the possibility of being wrong; ask questions rather than make accusations; identify and locate relevant documents; separate personal and professional concerns; assess their goals and seek and listen to advice. He also suggests a step by step procedure of obtaining and evaluating advice, deciding whether confidentiality can be maintained, keeping notes, obtaining support, deciding whether to blow the whistle and then determining an appropriate person or organisation with the power and resources to do something.

Bowie's (1982) requirements for justifiable whistleblowing include: appropriate moral motives to prevent unnecessary harm; using all available internal procedures before public disclosure if possible; having sufficient evidence; perceiving serious potential danger from the violation; acting in accordance with responsibilities in avoiding and/or exposing moral violations. Velasquez (1988) has combined questions from several authors as follows: how comprehensive and accurate is the information; what unethical practices are involved and why are they unethical; how significant and irreversible are the effects of these practices and are there compensating benefits; what is the obligation to disclose these practices internally or externally; and what will the likely effects and consequences be.

9. LEGAL PROTECTION IN THE US AND THE UK

Increasing awareness of the problems faced by whistleblowers in terms of loss of jobs, victimisation and other types of retaliation and their role, particularly in detecting and preventing fraud, has led to the development of whistleblower protection legislation. This section will concentrate on legal protection for whistleblowers in the UK and US, whre there are contrasting legal climates. The US has a mixture of state and federal legislation, as well as a political environment which (at least in theory) protects freedom of information and speech, but does not have comprehensive unjust dismissal legislation (Peritt, 1987), whereas the UK has legal protection against unjust dismissal, but a highly entrenched culture of secrecy (De Maria, 1997).

The US Congress established the Office of the Special Counsel in 1979, with the protection of employees from reprisals for protected activities, including whistleblowing, as one of its main purposes. Fong (1991) suggests that federal reprisal law has been complicated by two different and sometimes contradictory approaches to policy. One of these policies encourages disclosures, whereas the other promotes management's discretionary authority to sack workers to eliminate disruptions.

O'Leary (2000) highlights the ways in which the combination of federal and state laws in the US can often fail to protect whistleblowers. Federal protection is generally for the purpose of enforcing statutes which promote public welfare rather than in its own right, whereas comparable protection in state laws is about protection of the employee rather than support for other legislation. The federal Whistleblowers Protection Act of 1989 increases protection to federal government whistleblowers who disclose government waste, fraud and abuse of power. The first and fourteenth amendments to the US Constitution also give public sector whistleblowers some constitutional protection (Barnett 1992). However federal protection is limited, with some acts such as the Civil Service Reform Act and Whistleblowers Protection Act only applicable to federal employees, thereby giving most whistleblowers little protection at federal level. The main remedies for retaliation in federal whistleblower legislation are recovery of lost wages and benefits and reinstatement. Most statues allow court costs, including legal fees, and a few compensatory or exemplary damages (Poon, 1995). However federal statues are generally narrowly restricted by subject matter and applicability, whereas state statutory and common law generally provide wider protection (Poon, 1995).

More than half the US states now recognise the public policy exception to employment at will which allows employers to arbitrarily sack employees without contracts. This exception increases protection if the job losses are considered to be inconsistent with public policy and there is a broad definition of public policy. 37 states have general whistleblower statues (Benson et al, 1998), but only seventeen of them protect private sector as well as public sector employees (O'Learry, 2000). One of

the earliest was a 1981 Michigan act which prevents employers sacking, threatening or discriminating against an employee who reports a violation of federal, state or local laws, rules or regulations to a public body (Benson, 1992). There is also a great variation in the level of protection and available remedies in state legislation (Benson et al, 1998). The statues with the broadest coverage protect whistleblowers who make disclosures concerning suspected violations, mismanagement, gross waste, abuse of authority and threats to health and safety. A number of states require whistleblowers to first disclose complaints internally, but a number of federal and state laws require external whistleblowing (Poon, 1995). A summary of the employees covered and the designated reporting bodies under the different state whistleblower statues is given by Callahan et al (1994).

Most statutes define prohibited retaliation broadly and allow victims to file civil law suits, but may require them first to use administrative procedures. In most cases the burden is on whistleblowers to demonstrate that retaliation has occurred, though some statues put the burden on employers to show that their acts are unrelated to whistleblowing, sometimes by rebuttable presumption i.e. the assumption that negative workplace treatment such as relocation is in retaliation for whistleblowing if it occurs within a certain period after a public interest disclosure (De Maria et al, 1996; Westman, 1992). Remedies include reinstatement or recovery of damages, with only a few allowing punitive compensation and some providing for civil fines and other penalties (Barnett, 1992). The qui tam provisions of the US federal False Claims Act allow any citizen with knowledge of a fraud against the US government to bring a case in the name of the government and obtain 15 - 25% of the proceeds when the government intervenes and 25-30% otherwise (Raspanti et al 1998). Damages can be three times government losses and there is protection for employees who are discriminated against in any way as a result of a qui tam suit.

Dworkin et al (1998b) discuss the contradictory trends of the increasing use of secrecy clauses and increasing legislative and judicial encouragement to employees to blow the whistle. Although such nondisclosure agreements are only supposed to protect trade secrets and confidential information and should only be enforced if they are reasonable, they have been used to silence whistleblowers (Short, 1999) and journalists (Roberts, 1995). For instance a Michigan state court allowed General Motor to use a nondisclosure agreement to prevent a former engineer testifying about the dangers of fuel tank design in product liability suits. Brown and Williamson, the third largest US tobacco company, used nondisclosure agreements to obtain a temporary restraining order against a former company executive to prevent him disclosing information about the dangers of smoking cigarettes. Both rulings deprived the public of important health and safety information (Short, 1999). A TV company was forced to apologise and pay $15 million to two tobacco companies to avoid a $10 billion lawsuit after an investigative show had exposed an internal memo showing tobacco firms add higher doses of nicotine to make cigarettes more addictive (Roberts, 1995).

In the UK whistleblowers are now protected in certain circumstances by the Public Interest Disclosure Act 1998 (PIDA) (Lewis et al, 2001; Bowers et al, 1999). There are six specific categories of what are called qualifying disclosures and no 'other' category. However the legislation is complex and this complexity could possibly cause problems in implementation. Qualifying disclosures are those which tend to show that one or more of the following has occurred, is occurring or is likely to occur in the future: a criminal offence, failure to comply with legal obligations, a miscarriage of justice, danger to health and safety of any individual, environmental damage and deliberate concealment of information relating to the five previous categories (Bowers et al, 2001). An interesting question relates to protection for disclosures of offenses under international law. There is protection against victimisation and dismissal for making a protected disclosure is automatically unfair dismissal. However the degree of protection depends on whom the disclosure is made to, with the greatest protection for disclosure to an employer, and there is no right of action against any third party other than the employer who victimises the worker. When a worker has been victimised an award of compensation for the loss incurred can be made. Awards are assessed on the basis of what is 'just and equitable in the circumstances' but could be large, particularly if the whistleblower is unlikely to obtain another job.

Dworkin et al (1987) discuss state whistleblowing legislation in Michigan, Maine and Conneticut in some detail and consider the impact of the legislation in these three states. They conclude that the statues are no more effective in providing protection for whistleblowers or encouraging whistleblowing than the common law. Massengill et al (1989) again survey a number of early cases and conclude that there are restrictions on employer retaliation even when there is no specific protection. There is a clearly a need for investigations of the effectiveness of the legislation in protecting whistleblowers.

10. WHISTLEBLOWING IN THE SOVIET UNION

Lambert (1985) reports a study of 70 cases taken from the Soviet press between 1979 and 1983 and discusses eight cases in detail. There were 29 cases from the service sector, 12 cases about construction, nine about agriculture and one about a police officer. 80 offences are mentioned, including report padding

16 times, false reporting and other complaints about wages and bonuses 12 times, embezzlement 11 times plus four strong hints. In nine stories managers were accused of illegally using organisational resources for personal ends such as house building; and in six stories of using administrative control over goods and services to illicitly favour certain employees, particularly through control over the allocation of housing. Bribery was only mentioned four times, but this may be due to the fact it is relatively easy to conceal from outsiders and both bribe giver and taker are breaking the law. There were also many complaints about unfair dismissal and other disciplinary measures.

90% of the complaints were apparently personal, for instance about not receiving a flat or bonus or being dismissed, but could have been part of a wider social concern, whereas 10% were about general injustices, such as embezzlement, unjust distribution of flats or poor conditions at work. Lambert (1985) considers the group of disinterested protestors unusual in internalising the official ethic of responsibility and rejecting informal privilege and patronage. Although, in many ways supportive of these whistleblowers, press reports have various negative labels for them, such as 'quarrelsome', 'obdurate' and 'pedantic'. This labelling of whistleblowers is typical or reactions in, for instance, the US and UK where attention focuses on the whistleblower, who is labelled as the problem. Taking the official humanist and personal responsibility ethic seriously made these ethical whistleblowers outsiders. However Lambert considers that the fact that they are following the official ethic to be one of the reasons that complaints about abuse get so much coverage in the Soviet press.

As elsewhere, whistleblowers in the Soviet Union experienced retaliation. Despite apparently strong legal protection against arbitrary disciplinary measures or dismissal, managers did not find it difficult to get rid of employees who asked awkward questions. In 26 out of 66 cases the whistleblowers were dismissed at some stage of the conflict (although some were reinstated) and four left voluntarily, whereas 13 received some other form of penalty (Lambert, 1985). Management may retaliate, for instance, by setting up whispering and letter writing campaigns against critics or even trying to get criminal charges of slander brought against them or accusing them of blackmail.

Thus there are many similarities between the treatment of whistleblowers in the Soviet Union and, for instance, the US. In particular it is often the whistleblower rather than the 'offender' who is disciplined. However the underlying ideology and the mechanisms are different. In capitalist economies the profit motive gives rise to temptations to ignore safety standards and environmental costs and to pay low wages. In the Soviet Union the need to meet or if possible to exceed the various plans and the distribution problems and resulting deficits of goods and services led to temptations to bribery and report padding in order to obtain the materials necessary to fulfil plans or give the false impression they had been fulfilled. Therefore, though report padding occurs elsewhere, the need to fulfil targets probably made it more common in the Soviet context. Although breaches of law were not unconditionally tolerated (Lambert, 1985), there seems to have been some unofficial or even quasi-official acceptance of certain types of breaches of law in order to make the system function. There are again parallels in the US and UK, where, for instance, many firms in a number of industries fail to follow health and safety standards or dump chemicals (with toxicity beyond the official limits) into the environment, frequently without sanctions.

As elsewhere, Soviet whistleblowers were often isolated within the organisation, whereas the 'offender' often had some official support. However conflicts of loyalties may have not been as important as in, for instance, the US and the UK. Lambert (1985) considers whistleblowing to be part of long established and very widely used procedures for making individual petitions to state and social organisations, with the requirement of a prompt response with well argued decisions, including the grounds for any refusals. The party leaders encouraged citizens to act as watchdogs and blow the whistle on any illegal practices they discovered and promised protection from possible retaliation by management and local officials. Despite suspicion of informers, this may have lead to a cultural acceptance of whistleblowing, with loyalty to the party and its leaders or the wider society rather than the specific organisation.

11. HOW EFFECTIVE IS WHISTLEBLOWING?

In a number of cases whistleblowing has achieved at least some of its specific aims, but probably more rarely had an effect on public policy. Policy effects are generally difficult to determine, as there are often several contributory factors. However there is no research evidence on success rates, degree of success or compensation to vindicated whistleblowers. Therefore the fact that this section discusses 'successes' should not be interpreted as an indication of the overall effects of whistleblowing. Johnson et al (1990) report two cases of whistleblowing which did lead to policy change in the mid-80s and suggest that this was partly due to the status, credibility and political skills of the whistleblowers. They suggest that changes which are salient, specific and administratively feasible are more likely to succeed and that the particular political context with a Republican president and Democrat House of Congress may have also played a role. However other researchers (Soeken et al, 1987,

Trueslson, 1987) suggest that whistleblowing generally does not have a policy impact.

Hal Freeman, the regional manager of the San Francisco Office for Civil Rights in the US Department of Health and Human Services (HHS) resigned in protest at OCR policy discriminating again people with AIDS and related conditions. The subsequent publicity and group action amongst civil rights and gay activists led to a total change in policy. This would probably have happened eventually, but over a much longer time span if he had not resigned.

In the mid-80s Howard Kaufman, an influential professional in the Environmental Protection Agency (EPA) disclosed that the EPA was jeopardising public health by failing to enforce hazardous waste and toxic chemical laws and arranging deals with polluters and alleged that senior EPA personnel were failing to comply with environmental law on toxic and hazardous chemicals and misusing Superfund money. His allegations were repeated on a popular TV programme and he developed a coordinated strategy in association with two Congressional Representatives. Although he experienced reprisals, he was able to deflect efforts to silence him. His role in leaking critical memos and the investigation of his activities by the EPA, which drew further attention to EPA mismanagement, led to the replacement of the top leadership of the EPA. It also fed into increasing public concern about the risks of toxic chemicals and hazardous wastes. Hazardous waste law was strengthened and the Superfund programme expanded.

Whistleblowers in the US have also had a significant impact on the development of the anti-nuclear movement and public policy debate on nuclear energy (Bernstein et al, 1996). In the 1970s engineers at the US Atomic Energy Commission (AEC) leaked information to the Union of Concerned Scientists (UCS), indicating that the emergency core cooling system in nuclear reactors had failed major tests (Bernstein et al, 1996). AEC had suppressed these results to speed up the licensing of new plants. UCS published this information, leading to AEC hearings in 1972, at which the engineers reluctantly tested. The hearings, which lasted more than a year, contributed to the formation of a network that became the core of the anti-nuclear movement. After these hearings the focus changed from environmental issues to the threat of catastrophic accidents (Nelkin, 1971). Various bad construction practices became major issues, partly as many whistleblowers raised them at individual plants and attention shifted to the general competence of the utilities and the nuclear construction industry (Jasper, 1990). Three US engineers resigned from General Electric (GE) in 1976, due to concerns that GE was producing a reactor with known flaws, likely to cause a major accident. They helped focus criticism on generic design issues and contributed to creating public debate. They also drew attention to the willingness of regulators and the industry to suppress information and deceive the public (Bernstein et al, 1996). Whistleblowers have also been prominent in battles over particular nuclear plants. For instance the Government Accountability Project worked with local safe energy groups and supported whistleblowers and used their disclosures to help close two plants, Zimmer and Midland (Bernstein et al, 1996).

12. CONCLUSIONS

First they came for the Jews and I did not speak out - because I was not a Jew
⋮
and I did not speak out -
because I was not a trade unionist
Then they came for me
and there was no one left to speak for me
Pastor Niemoeller, victim of the Nazis in Germany

As this quotation indicates, whistleblowing is about speaking out. The consequences of not speaking out can be very grave, but speaking is not easy and generally involves a risk of retaliation, which could lead to loss of employment, relationships and mental health. However in some cases it may be appropriate, at least initially, to raise issues with the person concerned rather than publicly. Decisions to whistleblow generally involve balancing loyalties and duties and the likely consequences of both action and inaction. Although certain types of people may be more likely to whistleblow, whether whistleblowing occurs in a particular case probably depends on a combination of individual characteristics, organisational structures, the type and seriousness of the incidents and/or concerns and situational factors. There is an increasing body of legislation to protect whistleblowers. However, though there have been few scientific studies of its effectiveness, published reports indicate that many whistleblowers still experience retaliation.

Whistleblowing generally focuses on abuses within the system rather than challenging the nature of the system itself, which would require collective action by a union or workplace. Whistleblowers challenge fraud, health and safety violations, but not the nature of their organisation's activities. A number of companies, for instance, totally legally, sell weapons to regimes with bad human rights records, often to be used to suppress political opponents and human rights campaigners, as in the case of Indonesia. There is surely a contradiction in challenging fraud or safety violations, but not the basic abuses of such firms and the governments they support.

REFERENCES

AMA, (1997). *Developing a code of ethics in research:* Assoc. Americ. Medical Colleges, DC.

Anderson, M. et al (1980). *Divided Loyalties*, Purdue University Press.

Baab, D.A. & D.T. Ozar (1994) *JADA*, **125**, 199-205.

Barnett, T. (1992). *Labor Law J.*, 440-448.

Barnett, T. , D.S. Cochran and G.S. Taylor (1993). *J. of Business Ethics*, **12**, 127-136.

Barnett, T., K. Bass & G.Brown (1996). **15**, 1161-74.

Bayles, M.D. (1981). In: *Profess. Ethics*, Wadsworth.

Benson, G.C.S. (1992). *Manag Audit J.*, **7(2)**, 37-40.

Benson, J.A. and D.L. Ross (1998). Sundstrand. *J. of Business Ethics*, **17**, 1517-1527.

Bernstein, M. and J.M. Jasper (1996). Interests and credibility, *Soc. Sci. info.*, **35(3)**, 565-58

Bok (1981). Blowing the Whistle. In J. Fleshman (ed). *Public Duties*, Harvard University Press.

Bormann, E.G. (1975). *Discussion and Group Methods*, Harper and Row.

Bowers, J. J. Mitchell and J. Lewis (1999). *Whistleblowing,* Sweet and Maxwell.

Bowie, N. (1982). *Business Ethics*, Prentice Hall.

Bowman, J.S. (1983). Whistle blowing, *Public Admin. Review*, May/June. 271-276.

Braxton, J.M. and A.E. Bayer (1996). *Science, Technology and Human Value*, **21(2)**, 198-213.

Brock, J. (1999). *Symp Whistleblower Protection*, 497-529.

Caiden, G.E. and J.A. Truelson (1994). *Australian J. of Public Admin*, 53(4), 575-583.

Callahan E.S. and J.W. Collins (1992). *J. of Business Ethics*, **11**, 939-948.

Callahan, ES and T.M. Dworkin, (1994). *American Business Law J.*, **34(2)**, 151-184.

Chua, C.H. (1998). *Public Admin. Review*, **68(1)**, 1-7.

Clark, D. (1994). Whistleblowing. *Hong Kong Public Admin*, **3(1)**, 251-264.

Constanzo, P.R. and M.E. Shaw (1966). *Child Development*, **37**, 967-974.

CRI (1994). Commission on Research Integrity, Meeting 5. (Nov. 7), Office of Research Integrity.

De Maria, W. (1992). Queensland whistleblowing, *Australian J. of Social Issues*, **27(4)**, 248-261.

De Maria, W. (1997). *Crime, Law and Social Change*, **27**, 139-163.

De Maria, W. (1999) *Deadly Disclosures*, Wakefield Press.

De Maria, W. and C. Jan (1996). *Crime, Law and Social Change*, **24**, 151-166.

De Maria, W. and C. Jan (1997). Eating its own, *Australian J. of Social Issues*, **32(1)**, 37-59.

Devine, T. (1995). To ensure accountability, *The Scientist*, **9[10]**.

Drucker (1981). *The Public Interest*, **63**, 18-36.

Durso, T.W. (1996). Scientific dispute, not fraud. *The Scientist*, **10(4)**.

Dworkin, T.M. and M.S. Baucus (1998a), *J. of business ethics*, **17**, 1281-1298.

Dworkin T.M. and E.S. Callahan (1998b). Buying silence. *American Business Law J.*, **36**, 151-191.

Dworkin T.M. and J.P. Near (1987). *American Business Law J.*, **25**, 241-264.

EARC (1990), *Protection of Whistleblowers*, Electoral and Admin. Reform Commission Paper 90/110, Brisbane, Australia

Edsall, J.T. (1995). *Science and Eng. Ethics*, I, 329-340.

Edwards, S.D. (1996). *J. of Medical Ethics*, **22**, 90-94.

Elliston, F. J. Keenan, P. Lockhart and J. van Schaick (1985). *Whistleblowing Research*, Praeger.

Fong, B.D. (1991). US law, whistleblower protection and the office of the special counsel. *American Univ. Law Review*, **40(3)**.

Glazer, M.P. and P.M. Glazer (1989). *The Whistleblowers*. Basic Books, New York.

Gorta, A. and S. Forell (1995). Layers of decision: *Crime, Law and Social Change*, **23**, 315-343.

Goldbeck-Wood, S. (1997). *BMJ*, **315**, 1252-3.

Greenberger, D.B., M.P. Miceli and D.J. Cohen (1987). *J. of Business Ethics*, **6(10)**, 527-542.

Gunsalus, C.K. (1998a).*J. Sci. Eng. Ethics*, **4**, 51-69.

Gunsalus, C.K. (1998b). *J Sci. Eng. Ethics*, **4**, 75-94.

Hacker, A. (1978). *Across the Board*, **15**, 4-9,67.

Hoke, F. (1995). Novel application of federal law, *The Scientist*, **9(17).**

Jasper (1990). *Nuclear Politics*, Princeton University Press.

Jensen, J.V. (1987). *J. of business ethics*, **6**, 321-328.

Johnson, D.G. (2001). *J. Sci. Eng. Ethics*, **7(1)**, 151-2.

Johnson, R.A. and M.E. Kraft (1990). *The Western Political Quarterly*, 849-874.

Jos, P.H., M.E. Tompkins and S.W. Hays (1989). *Public Admin Review*, Nov/Dec, 552-561.

Judd, P.B. (1999). Whistleblowing, *J. of Business Ethics*, **21**, 77-94.

Keenan, J.P. (1995). *J. of Social Behavior and Personality*,**10(3)**, 571-584.

King, G. (1999). *J. of Business Ethics*, **20**, 315-236.

Klotz, I.M. (1998). *Life Sciences Forum*, 759-760.

Laframboise, H.L. (1991). Vile wretches and public heroes, *Canadian Public Admin*, **34(1)**, 73-77.

Lampert, N. (1985). *Whistleblowing in the Soviet Union*, Macmillan.

Larmer, R.A. (1992). *J. Business Ethics*, **11**, 125-8.

Leahy, P. (1978). The Whistleblowers, State Committee on Govern. Affairs, Washington DC

Lennane, K.J. (1993). *BMJ*, **307**, 667-670.

Lewis, D. and M. Spencer (2001) In*: Whistleblowing at Work*, D. Lewis (ed.), Athlone Press, 10-23.

Lindsay, RM., L.M. Lindsay and V.M. Irvine (1996). *J. of Business Ethics*, **15**, 393-407.

Lui, T.T. (1988). In: *The Hong Kong Civil Service and its Future*, I. Scott et al (eds.), OUP, 131-166.

Lubalin, J.S. and J.L. Matheson (1999). The fallout, *Sci. and Eng. Ethics*, **5**, 229-250.

McCutcheon, L.E. (2000). *Psychology*, 2-9.

McKnight, D.M. (1998). *Sci. Eng. Ethics*, **4**, 97-113.

Massengill, D. and D.J. Petersen (1989). *Employee Relations L.J.* , **15(1)**, 49-56.

Miceli, M.P. and J.P. Near (1984). *Academy of Management J.* , **27**, 687-705.

Miceli, M.P. and J.P. Near (1992). *Personnel Psychology*, **38**, 525-544.

Miethe, T.D. (1994). *Sociological Inquiry*, **64(3)**, 322-347.

Near, J.P. and T.C. Jensen (1983). *Work and Occupations*, **10(1)**, 3-28.

Near, J.P. and M.P. Miceli (1986). *J. applied Psychology*, **71**, 137-145.

Nelkin, D. (1971). *Nuclear Power and its Critics*. Cornell University Press.

OECD (2000). *Whistleblowing to Combat Corruption*, OECD Working Papers, 8(41).

Okie, S. (2001). A stand for scientific independence. *Washington Post*, August 5.

O'Leary, T. P. (2000). *Iowa Law Rev*, **85(2)**, 663-95.

Ontario (1986). Political Activity, Public Comment & Disclosure, Ontario Law Reform Commission.

Pamerlee, M., J.P. Near and T. Jensen (1982). *Admin. Science. Quarterly*, **27**, March. 17-34

Parrish, D.M. (1997). *JAMA*, **227(16)**, 1315-1319.

Parker, R.A. (1988), *Parliament. Affairs?*. 149-158.

Peritt, H. (1987). *Employee Dismissal, Bender.*

Perrucci, R. R.M. Anderson, D.E. Schendel and L.E. Trachtman (1980). *Soc. Problems*, **28(2)**, 149-64.

Peterson, J.C. and D. Farrell (1986). *Whistleblowing.* Kenndall/Hunt Publishing

Poon, P. (1995). *J. of Law, Medicine and Ethics*, **23**, 88-95.

Raspanti, M.S. and D.M. Laigaie (1998). *Temple law Review*, **71**, 23-52.

Roberts, J. (1995). Washington, *BMJ*, **311**.

Sci (1998). Science and Engineering Ethics, **4(1).**

Short, J.L (1999). Killing the messenger, *University Pittsburg Law Review*, **60(4),**

Sims, L.R. (2001). 'Sherry's secret' *Sci and Eng. Ethics*, **7**, 147-150.

Soeken, K. and D. Soeken (1987). *A Survey of whistleblowers*, Assoc. of Mental Health Specialities, Laurel, Md.

Sternberg, S. (1996). An unusual case. *The Scientist,* **10(1).**

Strader, K.D. (1993) *Univ. Cincinnati Law Review*, **62**, 713-764.

Truelson, J.A. (1987). *Corruption and Reform*, **2**, Spring, 55-74.

Tucker, D. (1995). Whistleblowing without tears, *Australian J. Public Admin*, **54**, 475-82.

Velasquez, M.E. (1988). *Business Ethics. Concepts and Cases*, 2nd edition, Prentice Hall.

Vinten, G. (1994a) In: *Whistleblowing Subversion or Corporate Citizenship*, G. Vinten (ed.), 3-20.

Vinten, G. (1994b). *J. Roy. Soc. Health*, Oct, 256-62.

Von Hippel, F. (1993). Russian whistleblower faces jail, *Bul. Atom. Scientists*, **49(2)**, 7-8.

Wenger, N.S., S.G. Korenman, R. Berk and S. Berry (1997). *J. of Invest. Medicine*, **45(6)**, 371-380.

Wenger, N.S., S.G. Korenman, R. Berk and H. Liu (1999). *Evaluation Review*, **23(5)**, 553-570.

Westin, A.F. (1981). *Whistle Blowing!* McGraw-Hill.

Westman, D. (1992). *Whistleblowing*, BNA Books.

Winfield, M. (1994). In: *Whistleblowing, Subversion or Corporate Citizenship*, G. Vinten (ed.), 21-32.

IFAC

Publications
www.elsevier.com/locate/ifac

THE FUNDAMENTALS OF MODERN CIVILIZATION CONSEQUENCES AND REMEDIES

Mohamed Mansour

*Automatic Control Laboratory ,Swiss Federal Institute of
Technology, ETH Zentrum,Pysikstr.3,CH-8092 Zuerich
Switzerland
email: mansour@aut.ee.ethz.ch
www.control.ethz.ch/info/people/mansour*

Abstract: Modern civilization depends on the rise of science in the 16th and 17th centuries followed by modern philosophy, which started by Descartes in the 17th century as well as the Illuminism ideas in the 18th century. From the different Philosophies of the last 4 centuries and considering which ideas have survived we can deduce four main parameters, which explain the history and the present situation. The dangers of recent globalisation and those of modern technology are discussed. Some directions of remedy are mentioned as a vision . *Copyright* ©*2001 IFAC*

Keywords: modern civilization, social science, ethics.

1. INTRODUCTION

I am dealing in this paper with modern civilization, which is essentially European and dominating in different parts of the world to different degrees. Like other civilizations it has its positive and negative aspects taking the value of live as basis which is surely relative.The higher the education level of the people the narrower the gap between the theory and the praxis of any civilization.

One may say that modern science (Russell, 1946; Henry, 1997; Tiwari, 1992) began in Europe when Copernicus (1473-1543) published his heliocentric theory in the year of his death in 1543 as hypotheses. All the astronomers as well as the church rejected it. Tycho Brahe (1546-1601) who adopted an intermediate position noted the positions of the planets for many years. His student Kepler who made use of these measurements discovered the three laws of planetary motion in 1609 and 1619.Galileo

(1564-1642) who is the founder of dynamics showed the importance of acceleration, discovered the law of inertia and the parallelogram law for projectiles and forces. He accepted the discoveries of the heliocentric system and used a telescope to discover 4 satellites of Jupiter. He was condemned by the inquisition privately in 1616 and publicly in 1633.Galileo renounced his ideas which caused a damage to the church. Newton (1642-1727) stated his three laws of motion, a part of which is due to Galileo and most important the law of universal gravitation. In the 17th century technology has developed rapidly with the invention of the microscope, the telescope, the thermometer, the barometer, the air pump, the improvements of the clocks, the magnet, Boyle law in chemistry and the discovery of the circulation of the blood. Mathematics has also developed: : Nabier (logarithms),Descartes(analytic geometry), Leibnitz (infinitesimal calculus).

13

Because of the triumph of science the relation between the humans and religion has shifted gradually in the direction of natural science and humans became not humble anymore. The causality principle became important. The 18th century witnessed three big developments namely the illuminism (Aufklaerung),the industrial revolution and the French revolution.

Modern philosophy (Diemer 1967; Russell, 1946; Bennett, 2001) was founded by Descartes (1596-1650) who considered a mechanical world and douting everything (Cartesian dout). Rationality is the way`` I *think, therefore I am``* And Knowledge of external things must be by the mind not by the senses. On the other hand Locke(1632-1704)did have another line of thought which is empiricism in the theory of knowledge. All our knowledge (with the possible exception of logic and mathematics) is derived from experience. In the line of Descartes is Spinoza (1634-1677), Leibnitz (1646-1716), Kant (1724-1804) and Hegel (1770-1831). In the line of Locke(1632-1704), Berkeley(1685-1753) and Hume (1711-1776).The two lines of philosophy are also reflected in ethics. Political philosophy has developed in two directions democracy and some sort of totalitarianism.

From the philosophy of the last 4 centuries I shall extract four parameters with which we should be able to understand history till today. The crisis of today is a good example of application. Also it is possible to make some predictions for the future and to give a vision for a better future of humanity. The four parameters are*: Belief in natural science, rationalism &empiricism (no metaphysics), seeking happiness as aim in life and ethics(no metaphysics).* These are the world view, the tools, the aim in life and the restrictions. These four parameters are extremes which may be softened by religious beliefs. Different Individuals have these parameters in different degrees. In the following four sections I shall give the consequences for the individuals and the society.

2. BELIEF IN NATURAL SCIENCE

This is a result of conflict with the religion. In mechanics the resultant is near to the larger force and in biology there is the survival of the fittest (Darwinism).Therefore the powerful gets what he can limited only by the law. For states, there is almost no international law which can be enforced, therefore there is no limit other than the powers of the other states. The result is war and economic injustice.

One can say that power has been the crucial factor in the history of mankind, but now it has been established according to the new belief (substitute

for religion). The force of fairness should play no part.

According to Russell (Russell, 1975*) the fundamental concept in social science is power with its many forms(wealth, military, civil authority, propaganda, secret service, priesty power).* Love of power is one of the strongest human motives even following the leader is trying to take part in power. Machiavelli glorifies naked power. Hobbes propagates that power of the state is absolute. Rousseau in his social contract tends to justify the totalitarian state and Nietzsche wants an international ruling race who are to be the lords of the earth. Belief in race and nationalism is natural outcome of love of power. Prophets have valued things other than power (wisdom, justice, universal love, …)

As Science is power then scientific activity is important for a society. For economic power work is important for the individual and society. The other side of it is the power of the stock exchange and currency market which can be misused . Information is a powerful tool in the hand of totalitarian systems. In democratic systems it plays the major role in the formation of the opinion of the people. It is normally misused and democracy is damaged. There is a saying may be not far from the truth (in a dictatorship the information is manipulated but the people know it and in a democracy the information is also manipulated but the people do not know).

Some consequences of the belief in natural science are colonialism, communism, fashism and the global dictatorship with its consequences like damage of the environment, political and economic injustice and international terrorism with its two parts as in fig.3.

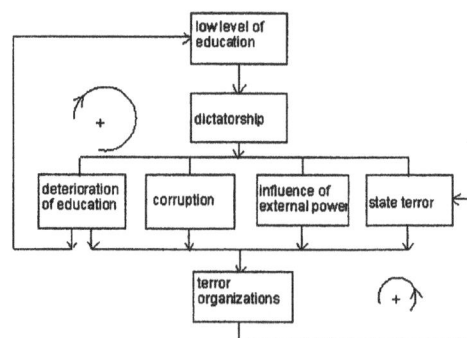

Fig.1 Dictatorship in a Developing Country

Fig.1 shows the cause-effect diagram of a dictatorship in a developing country (Mansour,2000), (Mansour,2001). The low level of education is a sufficient condition but not necessary for a dictatorship. Here we have two positive feedback loops which prevent the system from recovering. External interference is necessary for a considerable change towards a new development.

Unfortunately the modelled system is present in a large number of societies in the present world.

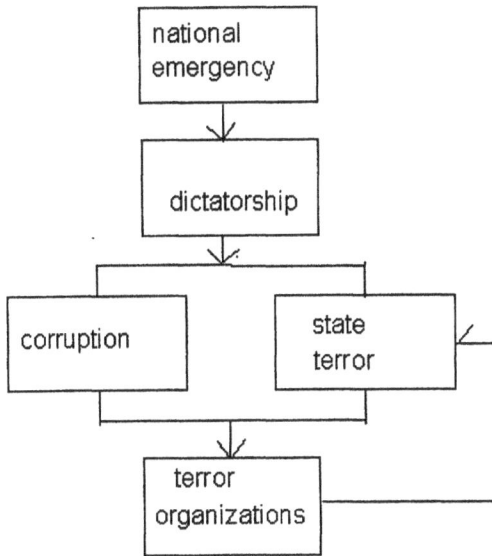

Fig.2 Dictatorship in a Developed Country

Fig.2 shows the cause-effect diagram of a dictatorship in a developed country. Here a national emergency is the cause of dictatorship which keeps its grip on power by misusing the information (propaganda). A terror positive feedback loop similar to that in fig.1 is present. A recovery can be from outside only. An example is the national socialist system in Germany.

Fig.3 shows the cause-effect diagram of globalization and the global dictatorship. This is similar to fig.2 except that the cause is the military and the economic power of one country (super power).A vision for the benefit of humanity is discussed at the end.

This system is similar to the dictatorship in a developed country. Here the members of the global system are countries of different system of government. The terror loop is quite similar. The united nations structure does not include a general assembly and an international court with binding decisions and the security council is not an international government with executive power controlled by the general assembly and the international court. That is to say, it is not a democratic system. The dangers of this system are the possibility of using weapons of mass destruction either by the global dictator or by the terrorist organizations.

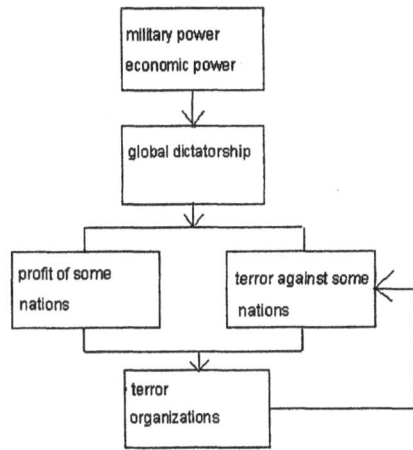

Fig.3 Global Dictatorship

3. RATIONALISM (MIND) & EMPIRICISM (EXPERIENCE)

Descartes:`` *I think, therefore I am* `` Kant ``the mind like the sun in the heliocentric system`` Spinoza, Leibnitz, Hegel and Marx follow the rationality whereas Locke, Berkeley and Hume follow the empiricism. Here we have no metaphysics. The consequences are: materialism, man is a machine and people like raw materials. Society forces individuals to instrumental reason where freedom is reduced. The economy growth is a main aim where the result is diverging capital and income in the society. The destruction of the environment, more technique in medicine with less human feeling as well as inhuman decisions are consequences.

4. SEEKING INDIVIDUAL HAPPINESS

Bentham (1748-...): each individual always follows what he believes to be his own happiness which leads to the happiness of the society ``greatest – happiness principle``. In economy, Adam Smith: do what is in your interest bounded by the law (interest of societies mostly not bounded by law). The consequences are: (Taylor, 1995)

A. Individualism: no high aim-loneliness-every one has his own values and should have nothing against the values of others which is moral subjectivity-concentration on himself-less interest in political activity thus loosing control of politicians which leads to mild form of despotism.
B. freedom: liberalism and respect of human rights in the society
C. democracy: Locke social contract-checks and balances- the legislative, executive and judicial functions of government should be kept separate is characteristic of liberalism

5. ETHICS

Ethics or the reference behaviour is determined by the mind (Kant), by the feelings (Hume) or according to utilitarianism (Bentham).

The model of Ethics is found in (Mansour 2001).

Ethics and legality: Fig.4 shows a symbolic representation of the regions of ethical and legal behaviour. The boundary of the region of legal behaviour coincides partly with the region of ethical behaviour. According to experience most of the behaviour is legal but not ethical according to the third parameter. Also the region of legal behaviour is expanding in a liberal society. Almost no ethics for international behaviour as interest is dictating.

Ethical problems in technology: dangers from nuclear, chemical and biological weapons from countries possessing them and from terror organizations. Also there are dangers of nuclear power plants for the people and the environment. Dangers of biotechnology come from the fact that experiments are done because of profit where the consequences are not clear.

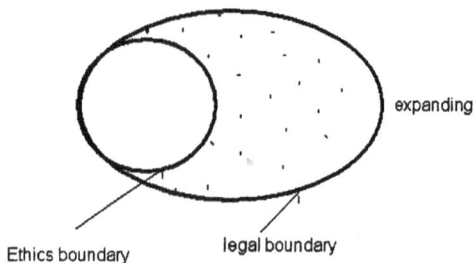

Fig.4 Ethical and Legal Behaviour

6. VISION FOR REMEDY

A. Belief in accountability (Spirituality). For most of the people there are no ethics without accountability. Thus making a correction to the illuminism.
B. Stressing common universal values or world-ethos (Kueng 1990). The project was presented to the United Nations General Assembly in November 2001.
C. Democratic world organization with an international government. `` When once an international government has been created, much of Locke political philosophy will again become applicable``(Russell, 1946)

What is a real democracy ?

I. decentralization to avoid accumulation of power
II. representation and direct democracy (Swiss model)
III. parties and independents in representation
IV. No concentration of information and building mechanisms to avoid misuse of information
V. Economics (no concentration of wealth, man is not a machine, adequate social security system)
VI. Education (general-all humans equal irrespective of colour, race, religion or world view-training in democracy-no racial history education)

D -Disarmament

7. CONCLUSION

In system theory one should have a model of the system under question, analyse its behaviour to understand it before making a synthesis. That is exactly what I have tried in this paper hopefully with some success. The situation of today is very crucial after the humans have developed the means of their destruction. *We have to live or die.* I have my personal experience with a dictatorship, global dictatorship and misuse of information. That means I am using the mind and experience in developing this work.

REFERENCES

Bennet, J. (2001). *Learning from six philosophers*. Clarendon Press, Oxford.
Diemer, A. and Frenzel, I. (Ed. 1967). *Phylosophie, Das Fischer Lexikon*, Fischer Buecherei, Frankfurt am Main.
Henry, J. (1997). *The scientific revolution and the origins of modern science*. Macmillan.
Kueng,H. (1990). Projekt Weltethos. Piper GmbH, Muenchen.
Mansour,M. (2000). System theory and human science. *Proceedings SWIIS 2000* (G.M.Dimirovski, Ed.), Pergamon Press , Oxford.
Mansour, M. (2001). System concept modelling in human systems. To appear in Kybernetes , MCB University Press, Bradford, England.
Russell, B. (1975). *Power*. Unwin Books, London.
Russell, B. (1946). History of western philosophy. London.
Taylor, C. (1995). *Das Unbehagen an der Moderne*. Suhrkamp, Frankfurt am Main.
Tiwari,S. (1992). *Mathematics in History, Culture, philosophy*. Clarendon Press, Oxford.

IFAC

Publications
www.elsevier.com/locate/ifac

PROVIDING AND R&D CAPABILITY FOR SMALL-MEDIUM SIZED FIRMS IN EUROPE: TOWARDS A UNIFIED MODEL OF TECHNOLOGY DEPLOYMENT, INNOVATION AND ORGANISATIONAL LEARNING

Larry Stapleton[1], Janko Cernetic[2], Donald MacLean[3], Robert Macintosh[3]

[1]Waterford Institute of Technology, Republic of Ireland
[2]Josef Stefan Institute, Slovenia
[3]University of Glasgow, Scotland

Abstract: Firm competitiveness today is intimately related to levels of innovation and the ability to produce and exploit new knowledge. Most EU firms face international competition on a truly global basis and operate in increasingly information-intensive environments. Set against this context, the paper addresses two related issues of fundamental importance to economic development: the competitiveness of indigenous small to medium sized enterprises (SMEs) and the development of an information society. *Copyright © 2001 IFAC*

Keywords: Advanced Technology, Innovation, Organisational Learning, Knowledge Production, Virtual Industrial Districts, Change Management

1. INTRODUCTION

Competitiveness relies on the ability to produce and exploit new knowledge. Large corporations tackle this issue by developing extensive R&D capabilities. This is not a solution available to most SMEs, however there are isolated cases where networks of small firms collaborate in order to share and exploit knowledge (the Italian fashion industry is one such example). Much economic development activity has concentrated on efforts to develop these industrial districts but it is argued here that the need for such networks to be geographically concentrated is being obviated by developments in new forms of ICT. Engineering and IT research must pioneer new forms of electronically mediated knowledge production which will foster collaboration between geographically dispersed groups of SMEs. By developing virtual industrial districts, this research offers the possibility for individual SMEs to tap into levels of R&D that none of the individual firms could sustain.

Effective knowledge production is the key to innovation and competitive success in today's information intensive environment (Gibbons et. al (1994)). Many large corporations deal with the issue of knowledge production through the use of dedicated R&D facilities and specialist external consultants, however such approaches are most often not an option for most Small to Medium Sized

Enterprises (SMEs). Nevertheless, research on the nature of knowledge production and early experimental successes may herald the arrival of a novel means by which small companies can compete effectively in the knowledge economy; the answer lies in the practice of "Mode 2 knowledge production", and in moving current mode2 models into electronically-mediated spaces.

2. TECHNOLOGY TAKE-UP IN SMES

Concerns about failures in the IT sector, including large scale business systems and group support systems, continue unabated (Stapleton (2000), Grudin 1994)). These concerns are also reflected in the work on social psychology and organisational learning (Postmes, Spears & Lee (2000), Argyris (1999)). It is recognised elsewhere that many modern information systems deployment projects are perceived by managers to be a significant risk to their business (Davenport (1998)) and may have a serious negative impact upon economic growth in the regions in which they are deployed (Byrne, Ryan and Stapleton (2000)). IS research and practise must deliver specific solutions which are grounded in specific application and problem contexts, rather than artificially imposed into organisational contexts with often traumatic results (Stapleton (2000), Cernetic & Jancev (2000)).

This paper argues that the poor take-up of ICTs amongst SMEs is driven by the fact that few small enterprises can clearly identify a business reason for adopting the technologies. This is borne out by recent studies such as SEISS (2001). Technology must make solid, business sense to SMEs that consider investing their meagre capital budgets into advanced systems.

The emphasis in Mode 2 approaches on contextually specific knowledge production offers unparalleled promise as a means of overcoming the problems outlined above. Under "e-mode 2", the issue is no longer knowledge or technology transfer; instead the issue is real-time production and consumption of knowledge in a given e-context, unhindered by the notion of transfer between different geographical locations.

3. TOWARDS ICTS WHICH ARE ATTRACTIVE TO SMES

In e-Mode2, technologies are deployed to mediate knowledge production, rather than as a store and retrieval mechanism for that knowledge. In this view IT is not primarily an information store and processor. Rather IT facilitates knowledge production, the generated knowledge becoming stored in the organisation itself. Here information storage and retrieval systems are only relevant in so far as they provide support for mode2 knowledge production. Applications that do not help groups produce knowledge become irrelevant. The focus in eMode2 is upon the interconnections and interactions of humans, rather than technologies. Emode2 recognises the nature of communications networks as proposed in Group Process theory. Brown (2000) tells us that

'The network of communication in a group is [a] crucial aspect of group structure. It is helpful to view communication channels in topological terms – as linkages – rather than in units of physical distance' (p.121).

Brown shows us that effective group structure requires an effective communications network in the group. Group processes do not necessarily require physical proximity to be effective. They DO require appropriate linkages in order to be effective. By extrapolation, mode2 can be effective as a group process if appropriate connections are put in place and managed, regardless of physical boundaries and limitations. This is not to say that mode2 will proceed in the same way as it would where players are physically proximal. Indeed, Ihde (1998)'s phenomenology of technology suggests that this is unlikely.

4. SUPPORTING KNOWLEDGE PRODUCTION IN SMES: CREATING THE VIRTUAL INDUSTRIAL DISTRICT

The Virtual Industrial District (VID) is a new concept introduced by this paper to describe a powerful model of economic development which focuses upon the specific knowledge-production needs of SMEs in an inter-regional context. The model is graphically presented in figure 1.

5. MODE 2 & ORGANISATIONAL LEARNING IN SMES

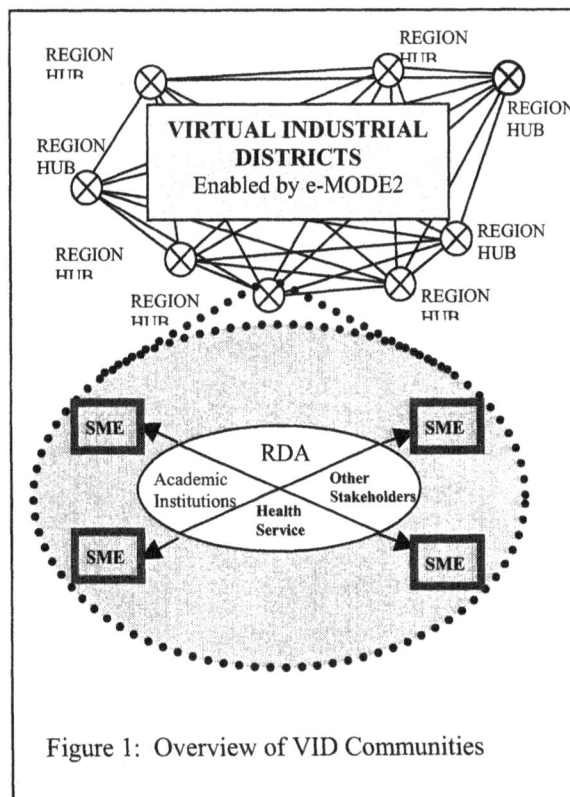

Figure 1: Overview of VID Communities

There is an established academic literature on organisational learning which can be traced back to seminal work by Argyris and Schon in the 1970s (see for example, Argyris and Schon, 1978). Many of these concepts were popularised by Peter Senge's book (1993.

The more recent material on Mode 2 knowledge production is entirely consistent with the sentiment that both individuals and organisations must learn collaboratively and collectively in order to compete. Our experiences with Mode 2 projects leads us to argue that only by creating and applying their own knowledge can organisations become truly competitive. This view has been developed over a series of projects with the SME community (see MacIntosh and MacLean, 1999 and 2001) and the public sector (see MacLean et al, 2001, Stewart et al, 2000).

In each of the projects which we have conducted, significant business benefits have been delivered to the organisations concerned, whilst academic insights have fuelled our research process. Some insights were transferred to the next project, but in each case, subsequent projects took prior knowledge and adapted, amended or abandoned it in the creation of a solution which would fit the new problem and context. It was this process which ensured relevancy in each case.

6. MODE 2 AND VIRTUAL INDUSTRIAL DISTRICTS

We have already described mode 2, as laid out by Gibbons et al, as a high involvement, problem solving, theory building dynamic involving academic institutions and commercial enterprises. We believe that there is scope to engage economic development agencies in the process in order to foster linkages between small, indigenous enterprises. These linkages may operate within one geographic area (as is the case with the Italian fashion industry example quoted earlier). However, we believe that by using electronically mediated communication, these linkages could span national boundaries in ways which were simply not possible until very recently. Such networks would not only provide for the development of new theory and academic knowledge; critically, it would enable the simultaneous creation and consumption of context-specific knowledge geared towards solving the problems and increasing the competitiveness of commercially active network members.

Using new communications technologies we can produce a cocktail which heralds the emergence of virtual industrial districts or e-clusters in which traditional barriers to technology transfer, innovation and competitive performance – such as geographical distance or physical mobility – are rendered meaningless.

7. ENABLING DEVELOPMENT IN OUTLIER & DISADVANTAGED REGIONS

The concepts presented here can help Europe to realise its potential as the most advanced information society on the planet. The framework harnesses the high levels of intellectual capital available in the Eastern and Western regions, in the development of a Pan-European knowledge society for the benefit of all citizens, independent of geographical location or technological advancement. The framework brings a new generation of user-friendly ICT-based services to regional development, which will have impact on industrial development of outlier regions and will

contribute to the reduction of regional and European R&D budgets.

The vision of regions in partnership is central to the ethos of this framework and is built upon the EU's principle of 'acquis communautaire'. It promotes and demonstrates the opportunities in Europe for high levels of innovation and economic development by leveraging off of the very cultural diversity that has so often been a hindrance to political, social and economic stability in the past. This proposal will address difficulties associated with EU enlargement as specified in Agenda 2000. To this end, this research proposed herein generates particular inputs to policy making at Community level and within Member and Associated States. Such inputs can be made available to Member States through the IST Programme Committee and to other interested parties in line with Article 19.3 of the Council Decision 1999/65/EC of 22 December 1998 on the Rules of Participation and Dissemination under Article 130j of the Treaty.

It is readily apparent that this approach focuses explicitly and directly upon the transformation of outlier regions into knowledge-production regions, which in turn creates opportunities within these regions for employment, industrial growth and increased competitiveness. The vision specifically includes many members of European society who are currently excluded due to geographical isolation or physical disability. This ensures that social-inclusion goals are satisfied as laid out in, for example, the *eEurope* initiative which aims at accelerating positive change in the Union and at bringing the benefits of the Information Society within the reach of all Europeans. This clear emphasis upon outlier regions in collaboration ensures that social cohesion is supported across Europe, and thereby contributes to political stability in the outlier regions. The research trajectory delivers large scale demonstrations and trials for the adoption and development of knowledge management technologies and services that involve citizens and businesses of all sizes across Europe in a knowledge production society. This also directly supports of the implementation of the *eEurope* action plan. Researchers are encouraged to focus efforts upon integrating the activities of European and regional development bodies, SMEs, other organizations in the supply chain as well as academic institutions who are traditional knowledge producers. The project has already received support and agreements to co-operate from many firms and agencies in the respective regions who recognize and endorse the important opportunity provided by this approach.

8. CONVERGENCE OF THIS MODEL WITH EXISTING MODELS OF ECONOMIC DEVELOPMENT IN POST-SOCIALIST COUNTRIES

This model of social and industrial development encourages Western-European and Balkan joint-partnerships in which cooperative knowledge production delivers industrial growth and economic development. Intercultural exchanges must occur at the level of individual organisational units in e-mode2 and happen in virtual space, eliminating geographical and time-zone barriers. E-Mode2 puts individuals in contact with each other, working towards mutually agreed aims and objectives, negotiated during the initial phase of e-mode 2 knowledge production. This model has been developed utilising concepts adopted in change management methodologies such as COPIS. COPIS has been successfully deployed in over 50 projects throughout the former Yugoslavia and addresses particular conditions in that arena (Cernetic & Jancev (2000), Jancev & Cernetic (2001)). Approaches like COPIS inform critical aspects of the e-mode2 process by addressing, in a post-socialist context:

1. Acquisition of Insight into essential business problems of post-socialist organisations
2. Identification of zones of innovation particular to post-socialist organisations
3. Creation of effective inter-cultural organisational teams, critical for virtual knowledge production

The original mode2 approach as utilised in outlier regions of Europe such as Scotland, already encompasses specific business imperatives found in outlier regions of western Europe. The E-Mode2 approach outlined in this paper focuses upon the specific business and academic issues facing all stakeholders in the mode2 knowledge production process, whether in post-socialist or western European settings. In this way e-mode2 can deliver a socially-accountable approach to knowledge production which is not culturally or economically imperialistic. This mitigates against technological imperialism and exploitation which often accompanies regional conflict or post-colonial settings and which remains too common in the discourse of regional development, particularly in the EU and other western blocs (Banerjee (2001), Faria & Guedes (2001)).

As Jancev & Cernetic (2001) point out, effective socio-economic change in the Balkans is only possible at a very fundamental level and, as Henning (2000) emphasises, effective innovation is most likely at intercultural interstices. E-mode2, as envisioned here, re-centres the change process within the social scene, rather than the technical scene,

recognising the unique social settings created by innovative technologies (Stapleton (2001)). The intercultural exchange required must happen in the context of economic recovery, where the practical problems people face are addressed through the intercultural exchange mechanisms. This will lead to the new outlooks and relationships required in this troubled region, and help create a 'rising tide which rises all boats'

9. CONCLUSION

This paper recognises that a key to economic and political stability in an expanded Europe is the development of a strong SME base in outlier and disadvantaged regions. This in turn requires that SMEs become major participants and benefactors in the, so-called, European Information Society.
However, evidence to date suggests that there is very poor deployment of ICTs amongst SMEs in European member states. At present there is little evidence that take-up of ICTs amongst SMEs in accession states is likely to be significantly different from the Northern and Western European counterparts.

This paper presents a vision of economic development which delivers tangible benefits to SMEs. These benefits include a new capability by which SMEs can produce knowledge. In this approach, SMEs are provided with an explicit and major R&D capability which, currently, is a luxury few SMEs enjoy.

Finally, this paper presents a vision in which SMEs cooperate in pan-European virtual communities in which shared business problems and issues are the binding agent. These are inclusive communities based upon mode-2 knowledge production concepts.

It is critical that European researchers pay great attention to the problems and opportunities associated with an expanded Europe. This paper attempts to forward debate by organising a high-level research vision and trajectory into a coherent framework. Political and economic stability is a tantalising prospect. However, without a major focus amongst social and technological systems researchers upon frameworks which help to resolve issues associated with European expansion, the promise of European Union expansion may turn out to be a hollow one.

REFERENCES

Argyris, C. & Schon, D.A. (1978). *Organisational Learning*, Addison-Wesley: Reading MA.

Banerjee, R. (2001). Biodiversity, Biotechnology & Intellectual Property Rights: Unpacking the Violence of 'Sustainable Development', *19th Standing Conference of Organisational*

Symbolism (SCOS XIX), Dublin, Ireland (forthcoming).

Brown, R. (2000). *Group Processes*, 2nd ed. Blackwell: Oxford.

Cernetic, J. & M. Jancev (2000) 'Implementation of Advanced Technology in Post-Socialist Countries', *Proceedings of the 7th Symposium on Automated Systems Based on Human Skill*, Brandt, D. and Cernetic, J. (eds.), International Federation of Automation and Control (IFAC), Elsevier: Amsterdam, pp. 251-254.

Byrne, S. & L. Stapleton (2001). The Illusion of Knowledge: The Relationship Between Large Scale IS Integration, Head Office Decisions and Organisational Trauma', *19th Standing Conference of Organisational Symbolism (SCOS XIX)*, Dublin, Ireland (forthcoming).

.Cernetic, J. & M. Jancev (2000) 'Implementation of Advanced Technology in Post-Socialist Countries', *Proceedings of the 7th Symposium on Automated Systems Based on Human Skill*, Brandt, D. and Cernetic, J. (eds.), International Federation of Automation and Control (IFAC), Elsevier: Amsterdam, pp. 251-254

Faria, A., A. Guedes (2001). 'In Search of Less Colonising International Management and Research: Bringing Suppressed and Surprising Experience from Academic Fronts', *19th Standing Conference of Organisational Symbolism (SCOS XIX)*, Dublin, Ireland (forthcoming).

Gibbons, M; Limoges, C; Nowotony, H; Schwartzman, S; Scott, P and Trow, M (1994), *The New Production of Knowledge: the dynamics of science and research in contemporary societies*, Sage, London.

Henning, K. (2000). The Aachen Region: Development through Information and Communication, *7th IFAC Symposium on Automation Systems Based on Human Skill: Joint Design of Technology and Organisation*, Brandt, D. & Cernetic, J. (eds.), VDI/VDE-GMA: Dusseldorf, pp. 1-4.

Jancev, M. & J. Cernetic (2001). A Socially Appropriate Approach for Managing Technological Change, *SWIIS 2001*, forthcoming.

MacIntosh, R and MacLean, D (1999), Conditioned Emergence: a dissipative structures approach to transformation. *Strategic Management Journal*, Volume 20, pp 297 - 316.

.MacIntosh, R and MacLean, D (2001) *Conditioned Emergence: researching change and changing research,* International Journal of Operations and Production Management, Volume 21, Number 9, forthcoming.

MacLean, D. M. Mayer, R MacIntosh and S Grant (2001), *Reflexive Strategizing*, Workshop on Micro Strategy and Strategizing, European Institute for Advanced Studies in Management, Brussels, 1-3 February 2001.

SEISS (2001). *Report of the South East Region Information Society*, February 2001, SEISS, Republic of Ireland.

Senge, P. (1993). *The Fifth Discpline: The Art and Practice of the Learning Organisation*, Doubleday: NY.

Stapleton, L. (2001). From Information Systems in Social Settings to Information Systems as Social Settings, *7th IFAC Symposium on Automation Systems Based on Human Skill: Joint Design of Technology and Organisation*, Brandt, D. & Cernetic, J. (eds.), VDI/VDE-GMA: Dusseldorf.

Stewart, G., MacLean, D. and MacIntosh R., *Applying Complexity Theory in Organisations (comparing experiences)*, in Complexity and Complex Systems in Industry, I P McCarthy and T Rakotobe-Joel (eds), The University of Warwick, 2000, pp 466 – 476

IFAC

Publications
www.elsevier.com/locate/ifac

TECHNOLOGY TRANSFER TO DEVELOPING COUNTRIES AND TECHNOLOGICAL DEVELOPMENT FOR SOCIAL STABILITY: PART 1

A. Talha Dinibutun[1] and Georgi M. Dimirovski[1 & 2]

1 DOGUS University, Faculty of Engineering
Acibadem, Zeamet Sk. 21, TR-81010 Kadikoy – Istanbul, Turkey
FAX#: ++ 90-216 / 327 96 31; E-mail: talhad@dogus.edu.tr

2 Inst. of Automation & Systems Engg., Faculty of Electrical Engineering
SS Cyril and Methodius University, MK-1000 Skopje, Rep. of Macedonia
E-mail: gdimirovski@dogus.edu.tr

Abstract: Part one of this paper identifies some crucial issues of the topic and develops a framework for a methodology of optimising development decisions in technology transfer for social stability and sustainable society development, which is the focus in the follow up second part. On the grounds of Professor Mansour's systems science approach and his philosophy of fairness and justice, and of ethics and ecumenical religious tolerance, it can be clamed that different standards and preconditions countries of so called Western Civilisation introduced insofar into tackling the issues of the globalisation are not justified in the eyes and consciousness of individual and collective anti-globalisation opposition. The issues and problems of a 'fairness' transfer of both knowledge and technology to the so-called Third World may well become a crucial tool for enhancing the positive effects of the globalisation. And, moreover, without this emphasis on the fairness and justice paradigm it seems no new world order can be achieved, but rather it may well turn into replica of some of the past empires with the new horror of sophisticated military. *Copyright © 2001 IFAC*

Keywords: Conflicting interests, development decisions, knowledge transfer, social stability, sustainable technology transfer.

1. INTRODUCTION

During last couple of decades, as the actual globalisation process as well as political and scientific discussion forums are expanding and quickening, the Mankind of this world is facing certain rapidly growing and expanding opposing civic reaction. Wherever political forums took place, everywhere wide anti-globalisation protest with international participation took place, and the last one in Italy turn out considerably tragic. This opposing civic reaction is not only internationalised but also more and more organised so that roomers about a new Anti-globalisation International are already in the air in many countries worldwide. Yet, it can be argued that the process of economic and other forms of worldwide integration is a historical stage of the development of the Mankind on this globe, and it is inevitably taking place. Why such an internationalised and growing opposing reaction is taking place?

Under the auspices of IFAC, its worldwide SWIIS community of systems scientists has social responsibility to search and find some part of the answer to this question, or rather a set of answers to issues involved. It may be argued with good reasons, a search to find an justifiable understanding on the globalisation process itself, and shed new light on the ways how to enhance the positive effects and suppress the negative ones, so that more and more people – individuals and nations – adopt some trust instead of fearsome opposition (e.g., see edited volumes by Kopacek, 1995; Dumitrache, 1997; Dimirovski, 2000) is indispensable.

The sustainable technology transfer for enhancing economy and society development in less and underdeveloped countries can provide some of the answers. For a longer period of time, some of the underlying issues of specific concern to developing countries have been explored through a series of events, dedicated to topics of Systems Approaches to Developing countries, by worldwide IFAC DECOM community (e.g., see edited volumes by Cuenod and Kahne, 1973; Hancke and Craig, 2000; Dimirovski, 2001). The science of systems and control has certainly the potential to find some of the missing answers and clarify the proper way for the future (Dinibütün and Dimirovski, 2001, Kile, 1995, 2000; Mansour, 2000).

2. WHY JOINT KNOWLEDGE AND TECHNOLOGY TRANSFER FOR SOCIAL STABILITY DEVELOPMENT?

Some may well find it inappropriate, however, the process of technology transfer is deep into the underlying problems arisen in the current globalisation process and its main controversy: Is technology transfer just a tool of the new form of colonising and colonialism or a tool of enhancing the economical advancement of the Third World and bridging the North-South gap? Some may well find it naïve, however, authors believe in the heart of the underlying problems arisen lie the issues of fairness and justice and realistically tangible equal opportunities for all in the meaning for all countries and nations and not solely for all individuals. In essence, it concerns the philosophy and the reality of fairness and justice as put forward by Mohamed Mansour in his plenary lecture at SWIIS 2000, in Ohrid, Macedonia. We believe, the continuing tragedies on the Balkans, Near East and elsewhere all lie in the heart of the globalisation process, also involving technology transfer and re-transfer, and more fair and justifiable re-distribution of wealth without which we can hardly

expect social stability and yet alone international social stability (Dinibütün et al., 2000; Mansour, 2000; Richardson, 2001; Scheffran, 2000).

A directly enhancing and supportive sustainable development approach using both knowledge and technology transfer as well as supporting the worldwide integration on equal opportunity basis for all nations may well be the proper effective way of achieving the convincing justification of the ongoing globalisation process. The real-life experience of the just elapsed century has demonstrated that, education, but also science as well as technology cannot remain neutral, no matter how and why they have been conceived. For, the motivation behind is tightly linked with competition and with tendency for prestigious position within the competition at best, and hence pre-conditioned by given human society and its features. This inevitably enhances conflicting interests, even crisis, unless in its systematic disposal for the benefit of mankind at large on a reasonable time scale is ensured. It is here where ethical categories of fairness and justice can play the role of primary incentives (Mansour, 2000; Dinibütün and Dimirovski, 2001).

It should be noted that a directly enhancing and supportive sustainable development approach using technology transfer as well as supporting the worldwide integration on equal opportunity basis for all nations precisely cannot be achieved without tightly inter-related joint technology and knowledge transfer to developing countries. Without fruitful knowledge transfer, the existing practice worldwide insofar has been producing some short-lived or dead-end intermediate advances (see, for instance, the relevant papers in Hancke and Craig, 2000, or recall the economic failure of most Far-East countries). Therefore these are prone to inefficiency or doomed to failure sooner or late.

Perhaps the best arguments on how important is the knowledge, and hence education, are recent, rather critical, reappraisal studies by Frank (2000) and Hanus (2000) concerning highly developed countries. On the other hand, the joint knowledge and technology transfer trough direct co-operation, and hence education enhancement, has demonstrated its developmental vitality (see, for instance, the relevant papers in Dimirovski, 2000). Of course, this alternative way of handing the problem of concern, among others, does require implementation of principles of fairness and justice (Mansour, 2000). To achieve this kind of situation to be a permanent, albeit evolving, process the technology transfer into less and underdeveloped countries must be accompanied with adequate sharing of knowledge in order to make technology transfer development enhancing further progress in such countries. Of course, practical implementations may take on various systematic procedures, and some of the recent ones emanating from medium-size Mid-European countries (Kovacz, 2000; Kopacek and Rommens, 2001) and from some other countries (Bloch, 2000; Richardson, 2001) may well serve as guidelines for this purpose.

3. PLACE AND ROLE OF TECHNOLOGIES BASED ON AUTOMATION AND CONTROL

All sorts of knowledge in the IFAC fields of expertise, no doubts, play a crucial role in technology transfer from the viewpoint of its essential role and impact of technology on the achievements of better efficiency, quality and productivity of industries and services within the competitive world of economy, national and trans-national (Savona and Jean, 1997). However, automated technologies are among the most expensive ones, and so is the knowledge associated, and moreover provide the cutting edge for the economy competitiveness thus not easy amenable to sharing. Therefore the principles of fairness and justice have to be implemented as incentives in both knowledge and technology transfer for the sake of common worldwide benefit on a longer run. Perhaps, we should make it clear, while enhancing the transfer of knowledge and technology that this process is costly in itself, and time consuming due to the sophistication involved. At the same time a considerable serious care must taken that transfer of automation and control technologies be directed solely to civilian sectors and human and social needs, that is the industrialisation (Bloch, 2000).

From the point of departure of the impact on enhancing the economical development, that is industry, these should be strongly focused on the low-cost automation in the infrastructures and towards rejuvenating the potential for quality and productivity by introducing more sophisticated control and automation in both production and service industries, in the first place. However, agriculture is also most important area of economy where to some automation technologies should be directed, because by and large developing countries as rule face the problem of food shortages.

Of course, already there are developing countries where advanced automation and control technologies are well under transfer process. For instance, in Turkey, South Africa and some Balkan countries high-tech applications like PLC technology, CNC machine based manufacturing, some industrial robotics and flexible manufacturing at early stage of developing

CIM and flexibly automated process technologies already pave their way trough. At the same time, the use of manually managed tools and/or process apparatuses is more usual typically. It may well be inferred that this is due to the higher level of general education and skills, and therefore clearly technology transfer is inseparable from knowledge transfer. Currently, the situation in some countries in Asia and Latin America also share similarities with the aforementioned one. Moreover, what is encouraging within this process of technological advancement in countries like these, a certain early stage of semi-automation in agricultural production is also gaining momentum. Nonetheless, all but most of the third-world countries the actual situation is not that encouraging at all (Hancke and Craig, 2000; Dimirovski, 2001).

In many of the developing countries the state is reached when companies can actual automate their manually managed machine tolls and/or process apparatuses at low or even not that low cost in order to have more competitive industrial plants. It is also important to note that, due to the respective globality feature, a great deal of communication, power and transportation technologies of today cannot function or even exist without a certain level of sophisticated automation, control and supervision, an management. For, even in the least developed of the developing countries, in order to achieve higher and competitive quality and productivity, the use of automatic control and other automated systems is unavoidable.

It seems rather important from the point of view of IFAC's cosmopolitan role, however, while taking the relevant care both with regard to the actual state-of-the-art in high-techs and with regard to speeding up a certain sustainable development in developing countries, that care is taken to increase the worldwide awareness on the usefulness and potential efficiency of low-cost automation systems. In particular, should the emphasis is put on obtaining better quality products/services and higher efficiency in using energy and raw materials, the awareness should be increased that all production and service technologies can be upgraded to higher competitiveness from mass application of low-cost control technology (Dinibütün and Dimirovski, 2001).

4. KAHNE'S WORK REVISITED: ON DEVELOPMENT DECISIONS IN KNOWLEDGE AND TECHONOLGY TRANSFER

Systems approaches to developmental problems in general do imply forecasting and planning, but require also relevant decision-making and optimising analysis,

and only then management execution. This is precisely the case with both transfer processes of technology and of knowledge, which cannot be properly accomplished on a day-today basis or solely following some intuitive decision policies. It is emphasises, however, that in doing so we have to confront the challenge of dealing with composite qualitative and quantitative factors as rigorous as feasible in order to achieve optimising development decisions (Gresford, 1972; Kahne, 1973).

It is this comprehensions that has led us to revisited Kahne's work and try to extended it further as to tackle joint knowledge and technology transfer problem of concern based on the background framework constructed in this and the previous paper (Dinibütün and Dimirovski, 2001) of ours. However, this is the topic of the future, part two of this paper, which deals with the relevant application of fuzzy system models and fuzzy algorithms. We end up the study insofar completed by citing two selected arguments, one by Gresford (1972) and the other by Kahne (1973), who initiated this line of thinking.

"One must stress that boundaries within which the problem are defined – especially complex and critical problems – are often unstructured, and quantification may not be readily possible. In dealing with qualitative problems or considerations, the means of taking into account qualitative values and putting them within a 'mix' of qualitative and quantitative variables requires much further effort on the part of management science community. Again, within a development context and in the light of policy options that presumably are the end result of a systems analysis effort, adequate means must be found to take qualitative factors into the analytic process." – G.B. Gesford

"The problem (i.e., making planning and development decisions – our emphasis) is separated into five distinct parts. Each part is discussed in the context of the planning process and each leads logically to the optimisation of development decisions. The five parts are goal definition, establishment of criteria, criteria weighting, alternative rating, and alternative ranking. The feature of making development decisions, which distinguished them from other optimisation problems, is what has been called 'fuzziness'. In any realistic problem formulation, the criteria are not precisely defined; they are fuzzy. The relative importance of each criterion is also fuzzy. Indeed, even one attempts

to rate a particular possible solution, he must deal with fuzzy information." – S. Kahne

5. CONCLUSION

The main issue of concern in here, as a matter of fact, is on how to create incentives and stimuli that enhance bettering the situation in developing countries for improving social stability according to the principles of fairness and justice, freedom and ethics, and to the international social responsibility. It has been explored in the context of recent advances in knowledge and technology transfer for enhancing sustainable development within the phenomenon of socio-economic and political globalisation. The further tackling of this problem and achieving some non-conservative yet applicable realistic results, however, has to be addressed via a methodology of systems analysis that combines qualitative and quantitative information through the use of fuzzy sets and system models with more its further. It is believed that this contribution sheds additional light on the main issue of concern as well as adds on some extensions to previous studies on assumptions for enhancing social stability, and also improving international stability, but much remains yet to be done in this systems analysis approach to optimising development decisions, which can significantly to overcome present deficiencies. This is the main topic of the future part two of this paper.

In this paper, we have argued and given evidence the technology transfer is inseparable from the knowledge transfer both for the reasons of pragmatic needs, and even more so because of the reasons of ethics and social responsibility of individual scientists and engineers, their professional societies and international associations of societies. The globalisation processes, no doubts, shall continue because Mankind's societies and national economies worldwide have come to that stage of development. Therefore, in the future, both the endeavours and the efforts of concerted activities of IFAC TCs on SWIIS, DECOM and EDCOM should converge so as to establish more effective actions focused on enhancing jointly both knowledge and technology transfer to all countries on this globe.

It should be pointed out, however, neither on national nor on international level the process knowledge and technology transfer shall deliver the expectations if it is left just to standard policy making and copying of existing experiences. System scientists and engineers as well as technology management specialist should enter on a much larger scale into the issues and problems involved. It is their social responsibility to elaborate models and methods for efficient yet most adequate conduct of knowledge and technology transfer well adjusted to the needs and potential in particular regions and even countries.

REFERENCES

Bloch, A. (2000). Technologies transfer to a developing country the road to industrialization. In: *Preprints of the 1st IFAC Conference on Technology Transfer in Developing Countries: Automation in Infrastructure Creation* (G.P. Hancke and I.K. Craig, eds.), Pretoria (South Africa), pp. 161-166. The IFAC and University of Pretoria, Pretoria.

Cuenod, M.A. and S. Kahne, Eds. (1973). *Systems Approaches to Developing Countries,* Algiers (Algeria). The IFAC and the ISA, Pittsburgh, Pennsylvania.

Dimirovski, G.M. (2000), Applied system analysis of transitional crisis of Southeast Europe and national goals formation. In: *Preprints of IFAC SWIIS 2000: Conflict Resolution in Regions of Long Confronted Nations* (G.M. Dimirovski, ed.), 22-24 May Ohrid (Macedonia), Paper ifac/SWIIS-09. The IFAC and The ETAI Society, Skopje.

Dinibutun, A.T., M.K. Vukobratovic, N. Shivarov, K. Schlacher and G.M. Dimirovski (2000), Roads to regional conflict resolution via academic and national goals of co-operation. In: *Preprints of IFAC SWIIS 2000: Conflict Resolution in Regions of Long Confronted Nations* (G.M. Dimirovski, ed.), 22-24 May Ohrid (Macedonia), Paper ifac/SWIIS-17. The IFAC and The ETAI Society, Skopje.

Dinibutun A.T. (2000). On the engineering education. In: *Proceedings of EPAC 2000: The 21st Century Education and Training in Automatic Control* (G.M. Dimirovski, ed.), 18-19 September Skopje (Macedonia), pp. 117-119. Institute of ASE, Faculty of Electrical Engg., SSs Cyril and Methodius University, Skopje.

Dumitrache, I., Ed. (1997). *Preprints of the 6th IFAC Conference on SWIIS,* Sinaia (Romania). The IFAC and Politecnica University, Bucharest.

Frank, P.M. (2000). Does the Green Card solve the problem of Germany's lack of engineers? In: *Proceedings of EPAC 2000: The 21st Century Education and Training in Automatic Control* (G.M. Dimirovski, ed.), 18-19 September Skopje

(Macedonia), pp. 77-80. Institute of ASE, Faculty of EE, SS Cyril and Methodius University, Skopje.

Gresford, G.B. (1972). Systems approach for development. *IEEE Transactions on Syst., Man & Cybern.,* **SMC-2**, 3, pp. 311-318.

Hancke, G.P., and I.K. Craig, Eds. (2000). *Preprints of the 1st IFAC Conference on Technology Transfer in Developing Countries: Automation in Infrastructure Creation,* Pretoria (South Africa). The IFAC and University of Pretoria, Pretoria.

Hanus, R. (2000), From Montaigne to Candide. In: *Proceedings of EPAC 2000: The 21st Century Education and Training in Automatic Control* (G.M. Dimirovski, ed.), 18-19 September Skopje (Macedonia), pp. 77-80. Institute of ASE, Faculty of EE, SS Cyril and Methodius University, Skopje.

Kahne, S. (1973). A procedure for optimizing development decisions. In: *Systems Approaches to Developing Countries,* (M.A. Cuenod and S. Kahne, eds.), Algiers (Algeria), pp. 385-392. The IFAC and the ISA, Pittsburgh, Pennsylvania.

Kile, F. (1995). A wellness model for of piece and stability. In: *Preprints of the 5th IFAC Conference on SWIIS* (P. Kopacek, ed.), Vienna (Austria), Paper FR-01. The IFAC and IHRT-TUW, Vienna.

Kile, F.O. (2000), The road to active piece (Key Note Lecture). In: *Preprints of IFAC SWIIS 2000: Conflict Resolution in Regions of Long Confronted Nations* (G.M. Dimirovski, Ed.), 22-24 May Ohrid (Macedonia), Paper ifac/SWIIS-02. The IFAC and The ETAI Society, Skopje.

Kopacek, P., Ed. (1995). *Preprints of the 5th IFAC Conference on SWIIS,* Vienna (Austria). The IFAC and IHRT-TUW, Vienna.

Kopacek, P. (2000), SWIIS – An important expression of IFAC's commitment to social responsibility (SWIIS 2000 Survey Paper). In: *Preprints of IFAC SWIIS 2000: Conflict Resolution in Regions of Long Confronted Nations* (G.M. Dimirovski, Ed.), 22-24 May Ohrid (Macedonia), Paper ifac/SWIIS-03. The IFAC and The ETAI Society, Skopje.

Kopacek, P. and E. Rommens (2001). Control technology transfer: Universities and SME's – the Austrian approach. In: *Preprints of the 2nd IFAC WS DECOM-TT 2001: Automatic Systems for Building the Infrastructure in Developing Countries* (G.M. Dimirovski, Ed.), 21-23 May Ohrid (Macedonia), Paper TT01-80. The IFAC and The ETAI Society, Skopje.

Kovacz, G.L. (2000). Technology transfer results achieved by Hungarian SMEs and academic institutions. In: *Preprints of the 1ˢᵗ IFAC Conference on Technology Transfer in Developing Countries: Automation in Infrastructure Creation,* Pretoria (South Africa), pp. 150-154. The IFAC and University of Pretoria, Pretoria.

Mansour, M. (2000), Systems theory and human science (Invited Plenary Lecture). In: *Preprints of IFAC SWIIS 2000: Conflict Resolution in Regions of Long Confronted Nations* (G.M. Dimirovski, Ed.), 22-24 May Ohrid (Macedonia), Paper ifac/SWIIS-01. The IFAC and The ETAI Society, Skopje.

Richardson J. (2001), The new petroleum pipeline-terminal project in Chad/Cameroon: Ethics, fairness and justice in the application of technology. In: *Preprints of the 2ⁿᵈ IFAC WS DECOM-TT 2001: Automatic Systems for Building*

the Infrastructure in Developing Countries (G.M. Dimirovski, Ed.), 21-23 May Ohrid (Macedonia), Paper TT01-27. The IFAC and The ETAI Society, Skopje.

Savona, P. and C. Jean (1997), Geoeconomica: Il dominio dello spazio economico. A cura di Paolo Savona et Carlo Jean, Franco Angely s.r.l., Milano.

Schffran, J. (2000), Power distribution, coalition formation, and multipolar stability in international systems: the case of Southeast Europe. In: *Preprints of IFAC SWIIS 2000: Conflict Resolution in Regions of Long Confronted Nations* (G.M. Dimirovski, Ed.), 22-24 May Ohrid (Macedonia), Paper ifac/SWIIS-07. The IFAC and The ETAI Society, Skopje.

B&SvP!

IFAC

Publications
www.elsevier.com/locate/ifac

TECHNIQUES OF PLANNING SETS OF MEASURES FOR PREVENTING AND OVERCOMING REASONS AND CONSEQUENCES OF EMERGENCY SITUATIONS

Kul'baV.V, Malyugin V.D., Shubin A.N.

Institute of Control Sciences of Russian Academy of Sciences
65, Profsoyuznaya str., Moscow, 117806, Russia
phone: (7-095) 334-90-09 fax: (7-095) 334-89-59
E-mail: kulba@ipu.rssi.ru

Abstract : General technology of planning comprehensive programs and measures for preventing and overcoming reasons and consequences of emergency situations (ES) is proposed, which is based on objective approach and aimed in a first instance at formulating goal set and implementing programs and planned decisions into specific procedures and activities which provide achieving and feasibility of formulated goals. This technology includes a sequential statement of goals, their supporting subgoals (tasks), standard sets of measures for their fulfillment, performance criteria and constraints, as well as their using to synthesize the optimal (rational) strategic plan for implementation of required sets of measures. *Copyright © 2001 IFAC*

Keywords : Emergency situation, complex measures planning

Objective-oriented programs differ by used goals and tools, they are multivariate in terms of purposes, and take a lot of money for their development and implementation. Under the limited resources it is necessary to integrate efforts and finances assigned for LOP elaboration in" terms of their comprehensive analysis and accounting interests of participating investors. In so doing one of the most important tasks will be a choice of a basic set of LOP from an assortment of objective-oriented programs, i.e. the set which has a financial support and takes a proper account of majority interests. So let us consider now formal models and techniques for LOP selection accounting their structure under limited finances.

Let $P = \{ p_i, i = \overline{1,I} \}$ be a set of programs aimed at preventing ES and providing a high degree of safety of some territory which is a confederation of sovereign states, and «interests» of their authorized representatives (AR) within the programs are differ-

ent. The members of a set $\Pi = \{ \pi_j, j = \overline{1,J} \}$ expose their interests through the amount of budget resources invested in wanted programs. Necessary expenses of implementing each discussed (proposed) program add up to S_i, $i = \overline{1,I}$. Within the assigned budget resources AR can invest the amount C_{ij} into the i-th program of interest. Assume also that programs to be implemented are being selected under limited finances R for the implementation of all wanted programs. So it is necessary to select the set of programs which would be acceptable for all parties in terms of the most complete accounting their own interests, as well as a fruitful and justified discussing by legal authorities.

To formalize the problem statement it is convenient to introduce a variable

$$X_i = \begin{cases} 1, & i\text{ - th program is included in a set for discussion,} \\ \\ 0, & \text{otherwise.} \end{cases}$$

Consider now the «interests» of different parties. Let

$$X_i = \begin{cases} 1, \text{if } S_i X_i - \sum_{j=1}^{J} C_{ij} X_i \leq 0, \\ . \\ 0, \text{otherwise.} \end{cases}$$

i.e. $Y_i = 1$ indicates that the i-th program is supplied with required investment, and, on the contrary, $Y_i = 0$ exposes the short supply of the i-th program implementation.

Obviously, the total deficiency over all programs selected for a discussion will comprise

$$f(X_i) = \sum_{i=1}^{I} \left(S_i X_i - \sum_{j=1}^{J} C_{ij} X_i \right)$$

(1)   Then $f(X_i) = \sum_{i=1}^{I} \alpha_i X_i$ is a finance deficiency function

(2) Y_i are the finance supply functions ($i = \overline{1,I}$)

Potentials of fitting and subsequent approving of LOP set appear to be significantly higher when

(3) $$max \sum_{i=1}^{I} Y_i$$

and

(4) $$min \sum_{i=1}^{I} \alpha_i X_i$$

are reached concurrently.

Introduce a matrix of individual states participation in LOP investment through their authorized persons as follows:

$A = \left\| \alpha_{ij} \right\|$, $i = \overline{1,I}$, $j = \overline{1,J}$.

where

$$\alpha_{ij} = \begin{cases} 1, & \text{if } C_{ij} > 0, \\ \\ 0, & \text{if } C_{ij} = 0. \end{cases}$$

($C_{ij} > 0$, i.e. the j-th state takes part in the i-th program investment. Then the function

(5)   Then $\phi_j(X) = \sum_{i=1}^{I} \alpha_{ij} Y_i$

will define a number of j-th representative's totally supplied programs and will express «interests» (gain) of each AR. While making his (or her) investment, every representative tends to reach

(6) $$max \sum_{i=1}^{I} \alpha_{ij} Y_i \ , \ j = \overline{1,J} .$$

It may be found by comparing (3) and (6) that objective functions are matched, i.e. overall interests are compatible with an interest of each individual AR. A further increasing j-th AR's totally supplied wanted programs is performed at

(7) $$\max_{(Y_i)} \sum_{j=1}^{J} \sum_{i=1}^{I} \alpha_{ij} Y_i \ , \ j = \overline{1,J} .$$

This increasing is profitable for both each AR and overall interests in a safety, so the program selection problem may be stated in the following form:

(8) $$min \sum_{i=1}^{I} \alpha_i X_i$$

under constraints:
— by each AR's budget resource

(9) $$\sum_{i=1}^{I} C_{ij} X_i \leq b, \ j = \overline{1,J} ,$$

— by the total cost of implementing all programs selected for a treatment

(10) $$\sum_{i=1}^{I} S_i X_i \leq R .$$

within the (8)-(10) problem statement that programs under consideration have the same priority. When some interstate body (center) exists which has centralized resources B_0 to support programs, it can take part in investment as a peer special representative Π_0 which has equal rights with other ones Π_j, $j = \overline{1,J}$.

At the same time the central body may have its own preferred programs which are believed to be the most efficient within the overall area under consideration. In this case, the center can invest resources B_0 into its priority programs with their subsequent selection for approval.

• Exact and approximate algorithms are available to solve the above stated optimization problem. In [0, 0,

0] it is interpreted as a problem of defining saturated arcs on a special type network.

• While planning and managing totalities of operations aimed at a prevention and overcoming ES consequences, it may be necessary to obtain appropriate data concerning the ES occurrence and progression, to counteract it effectively.

• One way of submitting information on the situation and produced responses involves using ES progression scripts, which may serve a major tool for effective decision making and coordination of response actions produced by a management system [0].

We call ES progression script a model of process which reflects environment changes in the context of occurrence and progressing ES in a discrete time space with a predefined time increment.

ES may be classified by their scale under two types, local and regional. Local scripts should be composed separately for each potentially dangerous object forming a basis for taking decisions by the management system (Object Commission or Staff).

Regional scripts are composed in the context of a totality of potentially dangerous objects within some territory and form a basis for taking decisions by the power bodies. Regional scripts include a list of potentially dangerous regional objects, their local scripts and associated counter measures in the event of ES occurrence and progression in the region.

To specify separate elements of a ES progression script, related concepts and models, the following definitions are introduced:

Definition 1. **Natural risk factor** is an undesirable state of the environment.

Definition 2. **Technogenic risk factor** is an undesirable state of the industrial object.

Earthquakes, floodings, mudflows are examples of natural risk factors, while toxic fallout, explosion, tire, radioactive pollution are the technogenic risk factors.

Definition 3. **ES event** is a realization of risk factors and counteractions of the opposite operative party.

Definition 4. **Situation** $S(t)$ at some instant of time t is a temporary orderable set of events happened before the instant t.

Definition 5. **Environment** $I(t)$ at some instant of time t is a *situation* $S(t)$ supplemented with information on available resources accessible by the operative party, on damages suffered, on expected delivery of resources and on taken management decisions.

Definition 6. **ES progression script** R is a process of changing *environment* within a discrete temporal space $R = R \{ I(t_i) | i = 0,1,...,u \}$.

Definition 7. **Script time increment** τ is the interval between two adjacent points of changing environment: $\tau = t_{i+1} - t_i$.

Time increment may be either fixed, i.e. $\tau_i = \tau = \text{invar}(i)$, or variable, i.e. $\tau_i = \tau_i(t_i)$, depending on a specific ES.

Definition 8. **ES reference time points** are the instants of time which are known from the past experience of a given type ES to be the points of strategically important events which result in dramatic changes of the *environment*.

Definition 9. **Reference script** R is a process of changing the *environment* as defined in ES *reference points*.

Definition 10. **Script fragment at the instant** t_i is a set of events and their interrelationships implemented in the interval from t_i to t_{i-1}. The script fragment defines situations which occur within the time increment $\tau = t_i - t_{i-1}$, including those at the instants t_{i-1} and t_i . It describes the ES progression environment during the time τ .

A *script fragment* allows to reflect a ES progression process within a specific time interval, perform the operative analysis of situations and their interrelationships, in order to make decisions on effective organization of counteractions, as well as produce and update operative plans of activities.

The script, as a tool for ES description and as its model, is bound to map situation occurrence processes and initiation of their interrelationships as a function of time. Sampling of time instants for estimating ES progression process and taking management decisions aimed at effective counteraction depend on a type and duration of ES, as well as on its anticipated consequences. A time increment r is required to match a real process of ES progression and be adequate to the interval within which new situations may occur causing new countermeasures.

By the management system (MS) operation mode, ES scripts fall into preventive (used for a day-to-day activity and higher readiness) and operative ones used in emergency mode. In the absence of preventive scripts express-scripts are used for overcoming emergency situations.

By the type' of probabilistic estimates of events related to ES occurring and progression, scripts are classified as pessimistic, basic (the most probable) and optimistic.

As a rule, the basic script of EC progression, being the most probable one, has a lower dimensionality and is convenient for a deep and careful study for the purpose to work out preventive and operative countermeasures. Pessimistic and optimistic versions of scripts may be emphasized also.

Pessimistic scripts of ES progression are sets of events and their interrelationships which result eventually in maximum losses and damages. Optimistic scripts, on the contrary, reflect events and their interrelationships which result in minimum losses and damages or eliminate them at all.

Local and regional scripts of ES progression are based on Risk Registration Certificates of potentially dangerous objects and of a region [0].
The Risk Registration Certificate of the enterprise (or objects) is a standard document that includes the following main sections:
General part,
Enterprise (Object) Structure,
Availability of Potentially Hazardous Materials,
Basic Type of Likely Crash,
Table of Possible Crash Interrelationships at the Enterprise,
Non-militarized Units of Civil Defense and Their Completeness,
Employment Security,
Distributed Evacuation,
Archival Incidents.

The following composition and content of ES progression scripts in different phases is proposed:

1. The likely interval of ES occurrence and progression or of its consequences within the time period $[0,T]$ is subdivided into discrete time instants t_i with a step τ .
The period T describes a time lapse for eliminating ES hitting seat and terminating further ES progression. Instants t_i describe control points of ES progression where corrective actions are applied for the purpose of improving environment and situation. The script step τ is selected so as to provide efficient action of available efforts, tools and resources during ES overcoming.

2. Formulating the initial environment (at t=0) and conditions resulting in occurrence and progression of ES, estimating losses and damages.
This phase includes modeling of ES occurrence and describing likely conditions of its progressing in the initial environment. Goals of counteraction are detailed, their efficiency is evaluated in terms of restricting ES progression and consequences, available resources are estimated. Preliminary pessimistic evaluation of losses and damages suffered from ES is performed. Countermeasures against ES progression in different versions are developed.

3. Establishing and putting into use quick response forces and tools to do salvaging and other urgent

works within hitting zones, depending on ES scale and damages suffered.

Procedure of quick response forces and tools mobilization and deployment within hitting zones is described, along with material and food supply, providing urgent medical care, as well as preparation for acceptance of additional counteraction forces and tools, including transportation facilities of all types. Hospitals deployment for victims rehabilitation are also described along with all measures, resources, efforts and facilities required for effective counteraction against ES.

4. Establishing current environment (at t_i) and conditions of ES progression, refining losses and damages estimates.
Current activities are described in this phase, similarly to p.2. The environment and the most important indices obtained at the instant t,-i are used as a current instant initial environment. Then countermeasures under new conditions are
defined, as typical for a given instant. The process is described in the same manner until assumed (or occurred) ES will be eliminated completely.

5. Checking a completeness and feasibility of obtained script and its updating for the purpose of maximum adequacy to the real ES progression. This checking is performed using expert knowledge through the learning and training under the developed script.

6. Establishing in the framework, of a given script a course of actions and operative plans of countermeasures against ES.
The methodological foundation of composing ES progression scripts is formed by techniques of structured and matrix analysis, as well as marked graph tools, allowing to separate out the ES seat structures, study their interrelationships and the entire ES progression process.

The formal methods of ordering and analyzing ES progression scripts, techniques for determining their main characteristics (say, mathematical expectation and variance of completion time both for individual tasks and the entire set of measures) are based on GERT network tools [0]. The organizational basis of composing ES progression scripts is formed by a management structure which aimed at providing safety and eliminating ES in the region.

In planning preventive measures it is convenient to use a special graph form of ES progression scripts involving cause relationship graphs (CRG) [0]. On the regional level, CRG vertices are the emergency events at regional objects, and arcs are causal relationships between them. Such a graph built for a set of objects, each being a potential source of occurrence and progressing ES of a specified type, is defined as graph G(S,B) with multiple sources { S^k, $k = \overline{1,K}$ } of ES.

A set of arcs B in a graph G maps processes of ES progression in the region, so interlocking an arc results in restricting or complete eliminating subsequent events in G.

Arcs are interlocked by allocating on them necessary and sufficient resources of different types, which provide a prevention or a restriction of ES progression processes.

As a result of G arc's interlocking, based on a rational allocating resources, forces and facilities available in a region, a subgraph G*= (S*,B*) of graph G will be obtained with a set of vertices S* which represent a set of events in the region, and set of arcs B* representing causal relationships between events which have not met with success to be terminated and will result in some damages and losses on occurrence.

Preventive measures against ES aimed at attaining the required safety level of a region may be represented by the orgraph Q= (D,L), where a set of vertices D map the measures (tasks) which provide meeting a desired goal, and arcs L are the interrelationships between measures defining the order of their implementation. Arcs of Q are oriented from source measures (tasks) on providing object safety to the desired ultimate goal represented by the vertex Do . Graph G is represented by the matrix Q=$\|q_{mn}\|$, $m,n = \overline{1,N}$, where q_{mn}=1 if n-th measure precedes m-th one, and q_{mn}=0, otherwise. Graph Q is related tightly to a source graph G(S,B) and based on its study. In so doing end events ac
hievable from the vertex So are separated in G, along with arcs connecting these events. For every such arc there are required measures D={Dm, $m = \overline{1,N}$ } which being implemented result in achieving the predefined goal Do. The integrated graph Q built in such a manner matches a set of preventive measures against ES, whose likely progression is defined by the given graph G. Interrelated graphs G and Q reflect essential features of the overall planning preventive measures against possible ES in the region.

The efficient preventive plan is formed on the base of optimal allocation of resources, forces and different type facilities required for implementation of planned measures to interlock the ES progression in a region under consideration as much as possible.

Formalize now the major efficiency criteria for building optimal preventive plans to guard against emergency situations. Depending on a type of ES in a region, the following objective function are proposed for use.

Likely human losses are defined by the function

(11) $\quad V_0 = \sum_{i=0}^{I}\sum_{j=i+1}^{I} v_{ij} x_{ij}$,

where v_{ij} — human losses from ES following an arc (i,j) and occurrence of the event V_j , x_{ij} =0, if the arc (i,j) is interlocked, and x_{ij} =1, otherwise. Likely damages are defined in a similar manner:

(12) $\quad U_0 = \sum_{i=0}^{I}\sum_{j=i+1}^{I} u_{ij} x_{ij}$,

where u_{ij} — damages expressed in terms of material values which would be destroyed under the events progression due to arc (i,j).

General cost consumed by putting available resources (forces and facilities) into operation, in context of implementing preventive measures against ES, may be defined as follows:

(13) $\quad C_0 = \sum_{i=0}^{I}\sum_{j=i+1}^{I} c_{ij} x_{ij}$,

Here c_{ij} — cost of transportation and putting into operation available resources (forces and facilities) of different types from points of dislocation to take measures on interlocking an arc (i,j):

$$c_{ij} = \sum_{m\in M_y}\sum_{f=1}^{F}\sum_{k=1}^{K} d_{kf}^{(i,j)} P_{mk}^{(i,j)}$$

where $P_{mk}^{(i,j)}$ — a required quantity of k-th type resources ($k = \overline{1,K}$) to implement m-th measure (a vertex of Q) on interlocking arc (i,j) in G, $d_{kf}^{(i,j)}$ — cost of transportation and putting into operation a unit of resources (forces and facilities) of k-th type f-th point of dislocation to interlock the arc (i,j) of the graph G.

In this context, the task of developing an optimal preventive plan of guarding against ES at the object, aimed at minimizing likely total damages within a region, will take the form:

(14) $\quad C_0 = \min_{\{x_{ij}\}} \sum_{i=0}^{I}\sum_{j=i+1}^{I} u_{ij} x_{ij}$,

under constraints:
— on a total reduced cost of resources (forces and facilities) allocated for implementing a set of preventing measures against ES in a region within a predefined period:

$$(15) \qquad c_{ij} = \sum_{i=0}^{I} \sum_{j=i+1}^{I} \sum_{m \in M_y} \sum_{k=1}^{I} a_k P_{mk}^{(i,j)} x_{ij} \le C,$$

where C — a total cost of resources allocated to a given region, and a_k — k-th type of resources allocated for implementing a set of preventing measures against ES in a region:

$$(16) \qquad \sum_{i=0}^{I} \sum_{j=i+1}^{I} \sum_{m \in M_y} \sum_{k=1}^{I} a_k P_{mk}^{(i,j)} x_{ij} \le R_k,$$

$$k = \overline{1,K},$$

where R_k — maximum quantity of k-th type resources (forces and facilities) allocated for the region,
— on a duration of implementing measures to interlock arc of G:

$$(17) \qquad \max_{x_{ij}} \left\{ T_{kp}^{(ij)} \right\} \le T_{given}, (i,j) \in B.$$

where $T_{kp}^{(ij)}$ the time to perform critical path activities on interlocking the arc (i,j), T_{kp} is the predefined maximum duration to complete all measures.

The problem (14)-(17) is a problem of integer linear programming which may be solved using standard techniques.

It is possible also to state problems of optimal planning preventive measures against ES in a region using the objective functions (11), (13).

REFERENCES

N.I. Arhipova, V.V. Kulba. Management under emergency situations (in Russian). M.:RGGU, 1994.-196 pp.

V.N. Burkov, G.Z. Kaziev, A.Z. Kuzmitzky, V.V. Kulba. Models and methods of goal programs selection (in Russian). M.: Nauka, Avtomatika i telemechanika. No 4, 1994.-pp. 135-142.

Kononov D. A., Kul'ba V. V, Shubin A. N. Stability of socioeconomic systems: Scenario Investigation methodology. — See this report.

D. Fillips, A. Garsiya-Dias. Techniques of Network Analysis (translated from English). M.: Mir Publishers, 1984. - 496 pp.

IFAC

Publications
www.elsevier.com/locate/ifac

APPLYING WORLD WATER ASSESSMENT TO CONFLICT RESOLUTION: 'MANAGING' NATURE TO EASE POLITICAL STRAIN

Jacques G. Richardson

Decision+Communication (Consultants)
Cidex 400, 91410 Authon la Plaine, France
decicomm62@aol.com

Abstract: Fresh water is a natural resource that is increasingly the direct cause of disputes, not only within countries, but also between countries and among regional groupings. International water planning and management are being handled by both public and private institutions and among a variety of United Nations units. The UN's new World Water Assessment Programme seeks to bring order to what had become a critically disparate situation—by creating system stability where it has barely existed. Project implementation should lead to a knowledge-management system concerning water, a working guidance group at global level for water planning and distribution, a worldwide knowledge-based system of consensus building/conflict management regarding water, and training programmes and materials for these four functions. *Copyright © 2001 IFAC*

Keywords: Natural resources; ecosystems, ecosystem management, control, system stability

Mini-glossary		
CIRBM	–	Collaborative Integrated River-Basin Management
CSWG	–	Case-Study Working Group
IHP	–	International Hydrological Programme
IWRM	–	Integrated Water Resources Management
PCCP	–	Potential Conflict to Cooperation Potential
WWAP	–	World Water Assessment Programme
WWDR	–	*World Water Development Report* (biennial)
Unesco	–	UN Educational, Scientific and Cultural Organiz'n

1. INTRODUCTION

The 2d World Water Forum (The Hague, March 2000) accepted as one of seven challenges the principle of shared water-resources management, now formally integrated for action within the United Nations' World Water Assessment Programme (WWAP). The seven challenges are the following:

Priority of legitimate demands

1 *Meeting basic needs* in water supply, sanitation and health
2 *Securing the food supply* via the equitable allocation of water for food
3 *Protecting ecosystems* including their integrity, sustainability

The "how to" of water-resources management

4 Sharing water resources *between different* users, different States
5 *Managing risks*: flood, drought, pollution
6 *Valuing water*, in terms of the economy, environment, culture and society (including careful, socially cushioned pricing)
7 *Governing water wisely*, meaning the involvement of good governance and actions by the stakeholders themselves.

The Programme's purpose is to resolve water-related conflicts, therefore, within an *integrated management of water resources*, abbreviated as IWRM. Action for this purpose is known within the UN as Potential Conflict to Cooperation Potential (PCCP), and the implementation of PCCP has been assigned to the International Hydrological Programme (IHP), managed and coordinated by the UN Educational, Scientific and Cultural Organization (Unesco), with headquarters in Paris.

Problems of 'water conflict' are those in water-resources management that involve at least two stakeholders (i.e. countries), each with diverging views on

- allocation,
- methods,
- standards, and
- objectives

that need to be considered, as well as the solutions applied in the course of implementing various stages of water-resources management. This definition is admittedly broad, and it places the entirety of water-resources management in the category of 'water conflicts.' Solutions to such management problems are always, therefore, compromises: effective tools for confidence building and conflict resolution.

2. OVERALL AIM

The main objective of this programme is to augment the cooperation potential concerning water, and capitalize upon it, while mitigating the risk that potential conflicts might turn into real ones. The instability of a political crisis (social scientists call this a *crisis instability*) is not very different from the mathematical concept of *chaos* (Saperstein, 1999). In terms of the variable intensity of different degrees of conflict, the programme may be interpreted by the following scale (IHP, 2001).

+4	+3	+2	+1	--	-1	-2	-3
Har-mony	Institut-ional mech-anism	Informal mech-anism	Tension	Diplo-matic action (neutral)	Open dispute	Armed conflict	War
+4	+3	+2	+1	--	-1	-2	-3

Table 1. Scale of intensities of conflict

Moving from left to right, the intensities marked "+" represent degrees of prevention, whereas *Diplomatic action* and the intensities marked "-" translate into the range of possible resolutions of a conflict.

The PCCP will have three tracks: by discipline, through education, and via case studies. The disciplinary track will consist of *legal means*; *historical experience and future analysis*; *negotiation, mediation, facilitation and public participation*; and *systemic analytical techniques* (models and formal procedures). The last group includes diplomatic approaches. The disciplinary track should yield state-of-the-art reports and records of achievement within the respective disciplines.

Education will be embodied by teaching materials (including textbooks), user manuals and tool kits, and Internet-based libraries/virtual references. Problem-identification teams will be associated with *disciplinary working groups*, who will advise the drafters of problem-solution output.

The case-study track will use questionnaires to facilitate analytical comparisons among the various conflicts dealt with. (An extract of a questionnaire appears on the last page of the present paper.) Large aquifers and water basins, whether national or international, considered conflictual by powerful water users—the stakeholders—as well as existing institutional mechanisms for conflict resolution, will figure among the primary criteria used in selecting case studies. Case studies will highlight established 'good practices' instead of analyses of 'what went wrong' and stress the validity of institutional function and IWRM—mentioned earlier, on the first page.

The number of case studies will depend on the availability of funding, voluntary participation by case-study agencies, and the level of coincidence with case studies associated with the conflict-resolution activities of the International Hydrological Programme. For the biennium 2001-2002, the following *water basins* are targeted as case studies.

Rhine River	Limopopo River (Africa)
Jordan River	Aral Sea (central Asia)
Mekong River	(southeast Asia)

3. METHODOLOGY ENVISAGED

Because case studies will be the concrete, utilitarian output of the analytical procedures, each study will be headed by a *task master* who will be assisted and supported by an *advisory case-study working group*. A questionnaire-based analysis will focus on (a) institutional efficiency and (b) cooperation potential in water-resoruce management in the fibe basins specified above. The educational track, already mentioned, will serve as simultaneous interface with current and future users of the PCCP output.

The educational trackwill be a simultaneous interface between current and future users of the PCCP output, particularly through reaching materials and educational modules. This track, led by a *track test team* (TTT), will represent the evolution of the respective project results as well as a 'drainage facility' for the two other tracks. The track will also establish a feedback mechanism, by way of pilot events and evaluation based on questionnaires, to yield educational modules, teaching materials, curricula and summer schools, and Internet-based dissemination. There will be two-week courses for decision makers and diplomats; summer schools will award Certificates in international water management. A reference library will be maintained on training and its tools. The tools required for decision support, e.g. negotiation support systems, will also be provided.

4. OPERATIONAL GOALS

As the three tracks are implemented and begin to interact with various IHP components, the PCCP concept will create a ripple effect by increasing the involvement of other UN entities sharing similar interests and aims. In the implementation of PCCP, there are five distinctly operational goals. These are
- defining and surveying conflicts in water-resources management
- monitoring indicators of *potential conflict* and *cooperation potential*
- developing educational materials
- providing tools for decision support
- disseminating results and encouraging good practices.

To realize these aims, a parallel approach to PCCP will engage the following actions.
- A *debate forum*, to address the basic conflict existing between *water for food* and *water and Nature* (at Bonn, December 2001)
- *Urging the study of initiatives* to use

conflict-resolution mechanisms; to advise PCCP on possible tools to be used during extreme events—e. g. floods—in regard to selected cases
• *Contact and cooperation* with other organizations and research projects related to PCCP thematically
• A *network to disseminate* expected PCCP results and good practices within CIRBM : the Collaborative Integrated River-Basin Management.

A specific goal is an open, internatonal conference on *PC* and *CP*, scheduled for 2003. This will serve as a forum to publicize on-going projects, research and other efforts in this area. Conference feedback should enhance the results of PCCP endeavours.   Output from the disciplinary, educational and case-study tracks will be processed by the PCCP Editorial Board—reviewing and putting into final form all pertinent PCCP 'production.'

The PCCP Editorial Board will be responsible for the following interlinkings:
• Preparing output for the World Water Assessment Programme (WWAP), and elaborating both PC and CP indicators as inputs to the biennial *World Water Development Report.*
• Evaluating debates initiated, managing networks collating data supplied by partners, and disseminating them appropriately.
• Reviewing the Board's own outputs and editing them for final emission.

These efforts will include the providing of educational material and training programmes; supplying decision-support tools; disseminating results and good practices; and providing PCCP inputs for the biennial *World Water Development Report.*

In addition to the above, selected systems will be prepared for demonstration at the 3d World Water Forum, scheduled for March 2003 in Kyoto.

5. CONCLUSIONS

Implementation of this project is expected to lead to the following results and positive practices:
▪ A worldwide knowledge-based system of consensus building and conflict management regarding water resources,
▪ Training programmes
▪ A system for communicating awareness of water-based problems
▪ A knowledge-management system
▪ A functional guidance group.

▪ As to the specific PCCP inputs to be prepared for the first edition of the *WWDR*

(2003), the programme will compile (a) a summary document relating to both the definition and the historical approach to conflictual problems, (b) a series of maps and other images highlighting the PCCP status concerning the major hydrological units of the world (basins and aquifers), and (c) a number of pertinent PCCP indicators.

An additional advantage, for the medium and long terms, should be to minimize the chances of terrorists seizing on water as a weapon to destabilize countries or alliances.

REFERENCES

IHP/Unesco (2001), "From Potential Conflict to Cooperation Potential" (Project document and rolling working-plan, of 2 April), Paris, p. 4.
Saperstein, A. N. (1999), *Dynamic Modeling of the Onset of War*, Singapore, World Scientific, pp. 3, 16, 18, 21 and *passim.*

See also
Beach, H. L., *et al.*. (Eds.) (2000), *Transboundary Freshwater Dispute Resolution*, Tokyo, United Nations University Press. Managing conflicts resulting from the quantity and quality problems of water round the world.

Biswas, A., and J. I. Uitto (Eds.) (2001), *Sustainable Development of the Ganges-Brahmaputra-Meghna Basins*, Tokyo, UN University Press. These basins contain the world's largest number of poor people inhabiting a single region.

Faruqui, N., *et al.* (Eds.) (2000), *Water Management in Islam*, Tokyo, UN University Press. Water is fast becoming the key development issue in the Middle East and North Africa.

Josselin, C., *et al.*, *L'eau/Water* (2001) (Bilingual guide), Paris, Ministry of Foreign Affairs and two additional governmental agencies. Treats the 'precious resource' thematically from Culture, through Development and Technology, to Social Life.

...

1.1.1.2 Political, social and economic characteristics of the basin

- Population, ethnicities, religions, languages, cultures

- Governmental regimes, political parties

- Brief human history, civilizations, wars, conflicts, general/human development indices, colonization

- Socioeconomic cross-sections, some key indicators, (e.g. main economic activities; regional, national and international interdependecies, GNP, per capita income, regional imbalances in income per country

• Political borders and their recent changes in context of:
ethnical borders or
territories—approximately
over the last 100 years, more if relevant to present situation

- Legal aspects (brief presentation of elements)

• Are there protected areas (national parks or reserves)? What are their legal and administrative status,
nationally and
internationally ?

1.1.1.3 International activities in the basin

• Have there been bilateral or multilateral cooperation projects active that are not internationally registered ?

- Have they collaborated with the internationally registered projects ?

- What have been their objectives and methodological approaches ?

- What have been the results on the IWRM and water-related conflicts ?

IFAC

Publications
www.elsevier.com/locate/ifac

INSIGHTS INTO FUTURE INTERNATIONAL SOCIAL STABILITY

Frederick Kile

Microtrend
420 East Sheffield Lane
Appleton, WI 54913-7181
USA
e-mail: 102610.2345@compuserve.com

Abstract: International social stability has typically taken one of three forms: 1) Expansion into relatively open regions; 2) Balance of power; 3) "Closed" periods during which internal factors minimized contact among separate national or ethnic societies. Future stability will be built on new approaches: working to solve the problem of the commons; learning from the future; education; demonstrating that excessive pride is counterproductive - in effect, a form of death wish; focusing on a balance between short-term and long-term stability to ensure a sustainable trajectory. *Copyright © 2001 IFAC*

Keywords: Environment, Future, International Stability, Peace, Social, Value Fields

1. INTRODUCTION

Earlier periods of international social stability depended on conditions which no longer prevail. Those forms of stability were based on: 1) Expansion into less-populated regions; 2) A balance of power; 3) "Closed" periods during which internal societal pressures minimized interaction with outside groups.

The bases for these earlier forms of stability have disappeared.

1. No remaining region can accept a large population influx without ensuing instability. Limitations include: space, infrastructure, and stressed ecosystems.
2. A "balance of power" depended on a combination of satisfaction and fear. A balance of power regime generally ended when one bloc felt strong enough to win a war with the other bloc. In a technological world, a major war could end civilization and might annihilate all human life.
3. The idea of a "closed" society seemed to have some validity until the Soviet Bloc collapsed. Few closed societies exist in the 21st Century.

2. EXAMINING SOCIAL STABILITY

2.1 Finding a Paradigm for Stability

A workable paradigm will be based on observation of how the larger society functions, including strengths and potentials for instability. This paradigm will acknowledge that differing "value fields" shape the attitudes and behaviors of individual societies. A value field is a broad set of interrelated values characterizing an individual, a group, or a society. Stability discussions will be productive to the extent that negotiators understand their own value fields and the value fields of other parties (Kile, 1984).

2.2 A Role for IFAC

IFAC engineers:
1. Understand how to control complex systems.
2. Have developed special types of logic to refine earlier control systems.
3. Understand the tradeoffs involved in maintaining system stability.
4. Will be able to incorporate value fields in models of international stability.

2.3 Examining Limits to Social Stability

In recent decades, social stability has been limited in several ways:

1. Human activity has breached environmental stability limits in some areas
   - Radiation release has made some areas uninhabitable.
   - Overgrazing has reduced food production in some regions, causing outmigration.
   - Mismanagement of water resources reduces environmental capacity and harms the population. Note population loss near the shrinking Aral Sea.
2. Unyielding political/religious positions destabilize relationships and cause conflict.

2.4 Preconditions for future Social Stability

There are at least two limits to international behavior:

1. War. In the new millennium, any war could escalate beyond intended bounds and cause mass destruction and huge loss of human life. War can no longer be justified — if war ever was justifiable.

3. Serious environmental degradation. When environmental factors become hostile, individuals and groups seek new niches. They move or change behaviors, postponing the effects of environmental degradation. Over time, the price of degradation cannot be postponed indefinitely. Groups eventually must adapt to a weakened environment.

2.5 Adapting to a Weakened Environment

Society does not solve social problems by overloading the environment. Example: Overfishing of North Atlantic cod brought cheap food for a long time — until the cod was threatened with extinction. Restrictions on fishing helped replenish the cod population. The lesson from oversfishing of cod might lead to wiser policies elsewhere, but that is uncertain. Depletion of the cod population caused serious economic dislocation in the Canadian Maritime provinces. These observations suggest that environmental integrity is integral to economic stability. Since economic stability is a foundation of political and social stability, environmental stability is a necessary condition for international stability.

2.6 Developing a Stability Paradigm

Historically, humanity has unintentionally limited its own growth and freedom through wars and stress on the environment. Both war and environmental stress are major constraints on society.

War and environmental stress have not been well controlled by agreements. Political compromises have not brought society beyond temporary stability because of shifting political exigencies. Moreover, environmental "compromise" is not a meaningful concept because the environment has no "voice."

Long term international stability cannot be built on reduction of tension through war or through offloading problems on the environment.

2.7 The Idea of a Future

Long term future stability will include aspects of voluntary self-limiting behavior. The concept of self-limiting behavior can be based on understanding that each society's value field includes an idea of a future. As each negotiator integrates his/her own idea of a future with the idea of a future brought to the discussion by other parties, the outcome will necessarily focus on a search for common ground.

Past stability regimes depended on circumstances no longer present or were a result of exhaustion following destructive wars. If past behaviors patterns are repeated, the likelihood of long-term stability is very low. Past patterns of international behavior resembled the "tooth and fang" law of the jungle, a behavioral model far too dangerous to function as a path to international stability. Stability depends on tacit or explicit agreements. It also depends on discussions which direct behaviors toward long-term stability, even if it is impossible to write and codified agreements describing expected international behavior.

As noted, environmental stress in some regions has exceeded the self-healing capacity of the environment. Once a certain threshold of environmental resilience is crossed, continued decline is likely, absent very costly intervention. Continuing environmental decline is clearly unsustainable in the long term.

When the public demands self-limiting behavior by its leaders, the public will establish a framework for social and international stability. Political leaders act when social pressure for action raises the political cost of not acting.

The public can demand voluntary self-limiting behavior by its leaders only when the public demonstrates that it can limit its own behavior in the interest of a future.

Two questions help to focus the issue: Is there any evidence from the history of war and peace that a strong stability paradigm has been found? Is the natural environment stable in its ability to support human activity?

In our opinion, the answer to both questions is, "No."

To have a future is to change. In the present era, to have a future will require that all parties re-examine their own ideas and change them if they are not consistent with a peaceful future. Continuation of present behaviors will almost certainly foreclose the idea of a future for the larger society.

2.8 Negative Sum Actions

It was once fashionable to speak of the futility of "zero-sum games." When groups clash, the result may be worse than a zero-sum outcome. A negative sum result can easily follow. Antagonists strive to gain a politico-military advantage, redirecting potentially peaceful efforts toward producing weapons — and often wars.

Similar negative outcomes follow as groups push the environment beyond its capacity to recover.

3. MOVING INTO A STABLE FUTURE

3.1 Beyond Negative Sum Actions

Attempts to win a "victory" in a competitive international situation can create a gain for one group. That gain, if any, is won at the expense of an even greater loss to another group. Military and economic conflicts damage the environment and often harm innocent neutral parties.

Groups often overlook the possibility of "negative-sum actions" largely because each party tends to think, "This situation is unlike anything which ever occurred before."

The well-known "Problem of the Commons" describes how seemingly innocent actions — taken to provide "something extra" for one party — create negative results for all.

3.2 Learning from the Future

Hegel wrote that people do not learn from history. It seems that Hegel was correct .

International stability will depend on people learning from the future. A vision of what might be may show leaders that cooperation is imperative for future survival.

3.3 Education and Dialog - Steps toward Stability

No one can be fully assured that education about future costs of present actions will stabilize group actions over long time periods. However, education can help leaders, and pressure groups which often drive larger groups, to soften motivations leading to negative-sum action.

Likewise, continuing dialog among parties competing for space and scarce aspects of the environment is essential to minimizing negative-sum actions.

Education must be offered in a spirit of ecumenicity. The spirit of ecumenicity is summarized in a statement, "We will do what we need to do in the short term to reduce the likelihood of disaster in the longer term." This approach can maintain stability in situations where even temporary instability would create chaos with little hope of re-establishing stability.

Pride is a major obstacle in the quest for future stability. National, ethnic, or individual pride is often a "death wish" in a competition for "victory." Rational arguments against irrational behaviors do not preclude irrational behavior.

3.4 Maintaining a Stable Trajectory

Given the obstacles to long-term stability and given
the reluctance of social groups to modify their goals
and behaviors, a sensible path to long-term stability
may be characterized by a continuing regime of short
term actions which keep the larger system within the
bounds of a stability trajectory.

REFERENCES

Kile, F. (1984) Cultural Factors and Values as
Influences on International Decision Making. In
*Preprints of the IX Triennial World Congress of
IFAC*, Paper 11.9-3, Vol. IV, pp 50-55.

IFAC
Publications
www.elsevier.com/locate/ifac

NEW TECHNOLOGIES IMPACTS ON DEVELOPMENT OF SOCIETY. ENVIRONMEMT, ROLES AND ACTORS.

Alexander Makarenko

National Technical University of Ukraine (KPI), Institute Applied System Analysis
Pobedy Avenue 36, 03056, Kiev, Ukraine

e-mail: makalex@mmsa.ntu-kpi.kiev.ua

Abstract: The global structure of society and the science as it subsystem are considered. Some new modelling principles are discussed and there consequences. The structures of science it poles and types are proposed. Interinfluence of society and science is discussed on the base of models with mentality accounting. *Copyright © 2001 IFAC*

Keywords: Global problems, Science and Technology, Actors of Evolution

1. INRODUCTION

No all recognises existing of challengers to the mankind and society in global development. The key factors in survival and sustainable development is new technologies. Besides direct impact on manufacturing the technology developments have a great influence on all regions of society life. All novelty in technologies is introduced through decision- makers. But who make the decisions? On first glance them are leaders, chiefs of large firms, international organisations and so on. But the situation is more involved. Mankind was developed in result of long evolution from primitive tribes. Such complexity follows to the situation when the leaders make solutions not arbitrary but on the base of accepted recently by society (or their part, for example by elite) norms, laws, beliefs and individual peculiarities of leaders. All issues above concerned also the future manufacturing, market and trade infrastructures.

Thus recent society is complex hierarchical object. Different individuals have different mental internal structures and play different roles. Because of many interconnections and hierarchical levels the whole society is very complex object and full modelling of them is unacceptable. Also under the question is the possibility of full modelling because evident presence of chance in processes. But recently new trends and prospects were proposed. Author hopes that his models (Makarenko, 1998, 2000) allow forwarding essentially simulation as for whole society as their parts and sub processes, including technology innovation.

We should especially remarked global problems: sustainable development, progress measure, global geoeconomics and geopolitics, information technologies, global education, nature evolution and impact of simulation on the future shaping of society. Some concepts of science as the whole are discussed.

Secondly we consider the different people inquiry to modelling in dependence on their role in society hierarchy and on the possibility of implementation. Another aspect of this problem is considering the new technologies influence on the individual in dependence of their role and personal mentality. Such problems are considered in connections with global education, information technologies and social psychology problems. The problems of social stability and plasticity under the technology impacts are discussed.

The author propose the review of existing modelling and simulation methods, their development prospects, their recent influence on the decision making and their role in future and some examples of applications (including Ukraine experience). We consider also the ethical problems connected with following from modelling power larger controllability of society.

The globalisation processes and the technical progress lead to the new type of an industry organisation – to the so-called Intelligent Manufacturing. Intelligent Manufacturing is only one of the problems closely connected to the Global World Manufacturing. First of all it should be remarked the notion of leading way of manufacturing or technological paradigm (Dosi, 1984; Glaziev, 1989; Lambin, 1993; Bye&Mounier, 1996). It is recognised that technological way is complex object which include space and time dimensions, hierarchical structure and social, political economical aspects. But the human factor and complexity of the World Manufacturing Regime as a whole object didn't received full description and understanding.

The topics above have direct implementations in technology transfer theory and practice. So in the report the author considers the approaches to the next problems: the models of the society as the collection of intelligent actors; the problems of education of organisation as the whole; the place of the leaders in context of connection of social design and self-organisation processes; structures and dangerous of global technologies. Some future research problems as the subject for collaboration is posed.

2. HIERARCHICAL STRUCTURE OF SOCIETY

Models (Short review from the author's papers). Let us described possible structures of models. Now we follow the description of hierarchical systems similarly the one in papers by Mesarovich and Takahara. We suppose that there are M hierarchical levels in the socio- economical system with N_j elements on j-th level. Each i- th element on j- th level have description by vector of parameters Q_i^j i=1,2,..., N_j ; j=1,2,..., M. Some elements on chosen levels can be in associations, marked by set of possible indexes in associations $L_i^j \subset \{1,2,...,N_j\}$. Many elements in developed society have a vast number of interconnections between levels. We may denote connections (bonds) between i1 elements on j1 level with i2 element on j2 level by J(i1,j1;i2,j2). Remark that another fields of interest (political, social, and educational) have similar network representation and society, as a whole is a union of such networks.

The bonds from the connection sets may be very different on the nature. The values of bonds may represent the normalisation of economical, informational, control channels, nationality, family bonds, and participation in professional associations and so on. The general model of system as in general system theory can be introduced with the help of input X1, X2, ..., XM and output Y1, Y2, ..., YM spaces for every level with input variables xi∈ Xi and output variables yk ∈ Yk.

In reality society are evolutionary system with dynamical changes on time. Further we for simplicity will consider only discrete time models with moments of time: 0,1,2,...., n,... .Following evolutionary nature of systems considered it is natural to consider as input of system in moment n the values of parameters from X in n-th time moment and as output the values at next (n+1) time moment (for n=0, 1, 2, ...). Remark that in developing society the content of elements set may changes. For example in economics the list of firms and corporations changes gradually by bankruptcy and by creating of coalitions. Social, political, governmental networks are often in transformations. This lead in general to changing the number of elements N_j (n) and number of hierarchical levels M(n) for different moments of time. Next if we wish to take into account the past states of society explicitly we should introduce to equation (1) or (2) the values X(0), X(1), X(2), ... , X((n-1)). Than the system description take the form Y(n)=f(X(0),X(1),X(2),...,X((N-1)), X(n), P, E).

Because we should consider evolutionary problems the main difficulties consist in searching the principles for modelling dynamics. The author's models consider the Society as large complex object constructed from many elements with interconnections. The considerations of Society properties allow picking out some interesting properties and then to propose the models, which can imitate society behaviour. Surprisingly the models are familiar with models of brain activity - the neuronets. Such models are under investigation by author since 1992 and yet had some interesting applications.

Now let us briefly describe the models. The first step of model building consists in the choice of model element and their description. Because it is need to take into account mentality of peoples in simplest models as the elements was took the individual with their description by series of mental and other (economical, demographical, and other parameters). These parameters may be evaluated in some scales psychology, sociology and other humanity sciences.

Next there are a lot of interconnections between elements in society - informational, business, relationship, infrastructure. The elements are connected by bounds. The bounds correspond to influence by individual, the money flows and others. Such connections are created historically. The set of element states and bounds give the description of society in some period of time. Remark that such description is familiar with verbal description in humanity sciences. For example the pictures in L.White's works remember the pictures for global socio- economical models. But if we wish to describe the dynamics of society and should to evaluate the influence of control than we must to know or dynamical laws or tendency in dynamics. The proposed models have such dynamical principles that them can imitate the behaviour of global culture in time. This is because the models have the property of associative memory. That is it can learns from historical processes the bounds and tends to very stable constructions- to so called attractor in pattern recognition in informatics and neuroscience. It is important that many social sub-processes in society also have the properties above allow to consider the separate sub-models (such as science).

3. SCIENCE AS COMPLEX SYSTEM.

Now we briefly pose the description of concept of science and technology followed from the works (Makarenko, 1998, 2000). The science is described as the complex subsystem of human society. Of course the development of elaborated models demand many

efforts but just the general principles of modelling give possibility to understand some regularities and yet to make some prognoses.

The science consists of system of scientific knowledge, infrastructure and organizations. The elements of scientific system are physical laws, researches, scientific organizations and their interconnections. This system has a long history of development and unique scientific system is stable structure with intrinsic internal regularities. Such approach allows many interesting conclusions. First is that we can consider the interaction of science with another subsystems of society (for example with mass media). The second is the different types of science organizations just like different culture type. Of course some intrinsic properties are common for all science types. But there may exist essential differences in sciences. Namely this aspect of models may follows some practical consequences for science in Europe.

Just like culture the science has multi- polar structure with different poles (centers). The poles may be in geographical space or in the space of scientific ideas and knowledge. From this point of view the structures of science in West and East Europe are the stable constructs (attractors) of large systems. Visual consequences of this intrinsic difference are vertical command- administrative science structure in East Europe countries and basically horizontal structure in West Europe countries. As the subsystem of society science has all drawback and disadvantages as former USSR society. The drawback of science in East Europe was their hierarchical building up (especially in USSR). This system absolutely ignored individual. Even now the scientific infrastructure in our countries doesn't directed to researchers. The scientific infrastructure "see" only the persons as chief of institute and higher in hierarchical structure. But mainly individual researches and small scientific groups receive the scientific results. This has one else consequence. The scientific results in East Europe countries frequently haven't restante. Existing vertical infrastructure also caused the difficulties in technological and social transformations flows from West to East (Vaseghy&Didenko, 1998).

The materials above concerned mainly the drawback of science infrastructure in East Europe. But nevertheless there are some interesting moments. Organisation of society in USSR followed to that science in USSR had specific way of thinking (mentality). This mentality was directed to more abstract problems and less to practical applications. Presumably such scientific style corresponds to some deep mental peculiarities on such territories. So for the purposes of enrichment of scientific methods it is optimal to combine reasonable scientific approaches from East and West.

Such model concept for science allows consideration of many new interesting problems. Now the technological and social innovations are recognised as one of the main source of long waves in industry. There are many investigations and models for such issues. But till now, as it seems for authors, there are a little model for the power of innovation flow. We proposed the general description of relations between the science as a whole system and industrial and social environment. The problem of dependence of innovation degree from investment into R&D and from the type of society is posed on strict background. This problem is closely connected to the practical problem of evaluation of the R&D part of budget of state or corporation. The next important practical problem is to define the best structure of network to technology transfer.

More elaborated models of proposed type is useful for understanding of life-cycle phenomena for industrial branches, commodities, society institutions. Preliminary investigations on our evolutionary models with time-dependent bonds discover the possibilities of sudden change of quasi- stable states with time. Such mechanism may be useful for understanding of concurrent replacing the commodities at market. formalisation. The example of practical importance problem is the effectiveness of different countries manufacturing ways. As example see different styles of industry (English-Saxon, continental and Japanese) and different goals of them (maximal profit, social defence, corporativity), (Jong, 1994). Also the self-organising criticality may be found in proposed models with real interpretation. It should be mentioned that now the observer's role and involvement of science into society operating are reconsidered leading to the entirely new research problems. The intelligent agent models with mentality accounting can solve such problems.

4. WORLD- SYSTEM THEORIES AND SCIENCE

The World- System theories have known in explicit form since the work of I.Wallerstein. Recently there are a vast number of theory variants (Hall, 1996). There are a lot of new problems of practical importance: core/periphery influence types of World Systems, evolution and cycles of development, global culture, origin of states. Our approach just in recent formulation can help in understanding such problems. And after development the more elaborated operational models we hope to receive the tools for more detailed investigations in considering society and technology

The attractors (attracting state) in our models should correspond to the World- Systems in the case of building the most general model of recent society as the whole. Then the core/periphery is the problem of introduction of control power for group of states in the core. Remark that familiar evaluation of power of poles had been done in our models of geopolitics. Evolutionary nature of Worlds- System may be considered in the models with changeable bounds between elements. Glokalisation (origin of local peculiarities under the background of globalization) corresponds presumably to origin of local attractors in our models. This is possible in the case of regions self-consistence. The changes of Worlds- System types may correspond to bifurcation in our nonlinear models.

Interesting phenomenon of practical interest (for prognosing goals) are the cycles of history, growth and failure of civilization, classification of historical periods, culture of society (Hall, 1996; Huntington, 1993; Toinby, 1990; Afonin, 2000; Lotman, 2000; Balonoff, 1987). In our approach such processes are natural phenomenon because of adaptive structure of models. Than the origin and dissipation of civilization is natural in our models. From this point of view the "end of history" anticipated in paper (Fukuyama, 1990) corresponds to existing one final attractor in models. Moreover the quasiperiodical behavior is intrinsic to our models in the case of asymmetrical bound. Also the chaotical behavior may exist in such case. Remark that asymmetry also present in the World – Systems. The examples are core/periphery asymmetry, asymmetry of informational, fuels and goods flows in recent world.

Frontiers, Cycles, Structures. One of the especially important phenomena is the frontier between different objects (as spatial as temporal). There are a lot of publications on such issues (see (Huntington, 1993; Hall&Jones, 2001; Dunaway, 1994), papers on cycles in time and structures. It is recognized (may be since Huntigton) that the frontiers between civilizations are very stable objects and such frontiers determine the geopolitical maps. Some frontiers are movable, with nonuniform properties (the examples see in (Dunaway, 1994; Hall&Jones, 2000). The borders of another type are considered in the problems of culture difference (Lotman, 2000; Harris&Moran, 1999). An examples are 576 potential types of frontiers, their change in historical problems, frontiers as complex object (Hall&Jones, 2001).

Our approach can help in introducing qualitative and quantitative understanding in frontier problems including different types of science. In our numerical results on geopolitical problems we received new frontiers between emergent blocks of states and hidden frontiers which can became real in some circumstances. Also in our approach the frontiers in different space scales may be considered as distributed zone where the elements from different states received new interconnections. It is evident in recent international relation (as example see recent Russian/Ukraine frontiers (StPt, 2001). Remark that the frontier may be subtler. An examples supply different cultures religious and ... types of scientific infrastructure. In our approach we can consider the real boundaries as the frontier between local attractors. Remark that collection of local attractors may constitute the global attractor (global state of the system) which surprisingly is described by mathematical theory of bundles. And at the end of this subsection we should mentioned that frontiers recognized as sources of new ideas, forms, development (creative zones).

Some Physical Analogies. As it was remarked in the introduction, one of the souses for our approach was the physics and synergetic. Our models incorporate ideas from them and may serve as the source of new ideas and research problems for physics. So in such context it is interesting to mention some recent development in the physical application to society investigation. First of all we wish to remark the discussion of some Wallerstein's ideas and physical concepts in (Prigogine, 2000). As it is usual for his works interesting all issues. In fact it contains the list of questions for any models: bifurcation, far from equilibrium conditions, sudden changes, complexity and structure, nonlinearity. But especially fresh are remarks on networks in society (including electronic networks and educational problems). Remark that ideas on dissipative structures now are one the key in any considerations of complex systems. In this connection we may remember also works on cities growth, capitalism as dissipative structure, works of Moscow school on nonlinear phenomenon (Kurdumov & Capitza, 1998, Prigogine, 2000). Proposed models are useful for implementation of above concepts. As was mentioned in the main text of the paper, now it exist new scientific field – econophysic. The author anticipates that proposed models can include the models proposed before. For example different kinds of networks (small worlds, power networks, society networks (Neumann, 1998; Mann, 1986)) may be considered as the examples in sub-networks in our models. As the consequences of physical principles we already discuss the punctuated equilibrium, self-organized criticality, instabilities, transitions, multivaluedness and dissipative structures.

5. MENTALITY ACCOUNTING

Agents with internal structures. The mentality accounting require considerations internal structures and incorporating them in global hierarchical models. The most natural way for implementing this task is to consider as model for internal structure also neuronet models. Remember that originally neuronet models were introduced in the investigation of brain. Firstly we can change the basic laws. On phenomenological level it may be implemented by introducing subdivision of elements parameters on external and internal variables and establishing separate laws for two blocks of parameters. But one of the most prospective ways for mentality account lies in searching equation also in neuronet class. Here proposed to introduce the intrinsic mental models of World in elements, which represent the individuals or decision-making organisations with human participation. The simplest way consist in representing image of World in the individual's brain or in model as collection of elements and bonds between elements. In such World pattern there exist place for representing individual himself with personal beliefs, skills, knowledge, preferences. Then the laws for evolution should depend on such representation.

Anticipatory property. The next step of developing models consists in accounting anticipatory aspects of individuals. It is evident that individuals in decision-making processes have prognoses on future. In such case the states of elements in model should depend on the images of the future described in internal representation. There may exist some stages of iteration in anticipating. We call such case as hiperincursion.

Now we give the possible structure of models and some corollary. First we describe the model structure with one element with internal structure. If there were no internal structure it were the system in section above for dynamical law. Let the individual with internal structure has the index $i=1$. Their dynamic is determined by two components. First component determines by external mean field as above. Second part of dynamic is connected with internal dynamics of first individual. Remark that this dynamic partially account the willing of individual. There exist many models for such part of dynamics but it is useful to put the neuronet models for our purposes.

Let us named the pattern of society $Q^{(1)}(t)$ in section above as 'image of real world ' in discrete moment of time t. We also introduce the $Q_{wish}(t)$ - ' desirable image of world in moment t by first individual' as the set of element states and bonds wishes by first individual in moment t. $Q^{(1)}_{wish}(t)=(\{s_I^{wish}(t)\}, J_{ij}^{wish}(t)\})$, where s-states of elements and J - values of bonds. Then we assume that the change of first individual state depend on difference between real and desirable image of the world. The resulting system takes the form: $S_I(t+1)=G_I(\{s_I(t)\},\{s_I(t+1)\},\ldots,\{s_I(t+g(I))\}, R)$, where R the set of remaining parameters. It is very prospect that the structure of system above coincide with anticipatory systems with incursion. This follows possible similarity in properties. Such description and models allow consider problems connected with technological skills, initial and high education, ethics and moral.

SUMMARY

Thus proposed models allow considering problems of interrelation society/science. As example we may mention sustainable development and investment into technologies and science (preventing possible collapse of society structure), new background for considering diffusion of innovation, global education models and many others. Proposed consideration may also serve as the source of practical consequences for different international organisations. One of such recommendation is need in protection of Science types diversity. (Remember that we try to save the biological diversity). Such investigations may be implemented in international research projects. As example for considering it may proposed the next: 1) S&T global models development; 2) Impact of different scenarios of S&T policy on society stability; 3) Optimisation problems formulation; 4) Global and local aspects of science impact and practical consequences

REFERENCES

Afonin E., Bandurka A., Martynov A. (2000) Social development AD. Kyiv, Parliamentarian Publishing House. 310 p.

Balonoff, P. (1987) The mathematical theory of culture, Vienna, Austrian Soc. of Cybernetics.

Bye P.&Mounier A., (1996) Growth patterns and the history of industrialisation. Int.Social Science Journal, N.150 (December). Pp. 536-549.

Dosi G., (1984) Technological Paradigms and technological trajectory. In. Long waves in world economy. L.

Dunaway, W. (1994) The Southern Fur Trade and the Incorporation of Southern Appalachia into the World- Economy, 1690- 1763. Review of Fernand Braudel Center, **Vol.18**, pp.215-242.

Fukuyama, F. (1992) The end of History and the Last Man. N.Y., Toronto, 272pp.

Glasiev S., (1989) Long-term aspects of innovative policy (On the 'Long-waves' in technical-economical development). Moscow, Proc. Inst.of System Research, P.4-25. (in Russian)

Hall, T.D. (1996) World- Systems and the Evolution. An Appraisal. J.World. Syst.Res., **vol.2, n.4.** pp.1-30.

Hall, T.D.&Jones, L.M. (2000) Comparing Frontiers: Approaches from World- System Analysis. Panel on Regional Studies and World History/Global Studies, at Paradigms in World History: Global Studies and World History, Binghamton Univ, March 3, 2000. 14 p.

Huntington, S. (1993) The Clash of Civilisation? Foreign Affairs, **Vol.72, n.3**. Pp.22- 29

Harris, Ph.&Moran, R. (1999) Managing cultural differences. Gulf Publishers Company,

Jong de X.B., (1994) European capitalism: between freedom and social justice. Voprosy Economiki. N.4. pp.106-109. (in Russian).

Kurdumov, S.&Capitza, S. (1998) Books on nonlinear systems, Moscow, (in russian).

Lambin J.- J., (1994) Le Marketing Strategique. Ediscience Int, Paris.

Lotman, Yu.,M. (2000) Semiosphere, St-Peterb., Iscusstvo, 740 p. (in Russian).

Makarenko A. (1998) New Neuronet Models of Global Socio- Economical Processes. In "Gaming /Simulation for Policy Development and Organisational Change" (J.Geurts, C.Joldersma, E.Roelofs eds.), Tillburg Univ. Press, 1998. pp.133-138.

Makarenko A. (2000) Models with anticipatory property for large socio-economical systems. Proc. 16 th World Congress of IMACS, Lausanna, 21-25 August, Switzerland, Paper n. 422-1

Mann, M. (1986) The Sources of Social Power: A History of Power from Beginning to AD 1760. Cambr. Univ.Press.

Neumann, P. (1998) The small worlds. Santa-Fe Institute Working Paper, Santa- Fe, Calif, (USA).

Prigogine, I. (2000) The Networked Society. J.World-Syst.Res, **vol.6, n.3**. pp. 892-898.

StPt. (2001) Proc. Conf. Geopolitical and Geoeconomical Problems in Russia- Ukraine relations (values, prognoses, scenarios). Ed.D.Nicolaenko, St- Petersb. Russia, 22-24 January 2001. (in Russian). [www.geography.net.ru].

Toinby, A. (1991) The comprehending the history. Moscow, Progress. (in Russian). 736p.

Vaseghy S.&Didenko N., (1998) Electronic message from EUROSCIENCE WG on Technology Transfer on 16 July 1998.

IFAC
Publications
www.elsevier.com/locate/ifac

SOME ASPECTS OF TECHNOLOGY TRANSFER –
A CASE STUDY

Antoni Izworski, Jozef B. Lewoc and Bronislaw Piwowar

Wroclaw Technical University, Wyb. Wyspianskiego 27, 50-370 Wroclaw, Poland
BPBiT Leader (Leading Designer), ul Powst. Sl. 193/28, 53-138 Wroclaw, Poland
Networld Poland, ul. Tytoniowa 20, 04-288 Warszawa, Poland

Abstract: The paper presents some technical and social aspects of the technology
transfer on the example of the Polish biggest ex-manufacturer of computers – Elwro.
The history of Elwro is followed at its milestones – the launching of its leading
products (computers) and the fall-down of the enterprise. Major technical and social
aspects are discussed for each milestone. 2001 is the year in which the IFAC
meeting was held. *Copyright © 2001 IFAC*

Keywords: information technology, transfer, social aspects, technical aspects,
computer manufacturing, computing systems

1. INTRODUCTION

The Wrocław Electronic Enterprise Elwro,
established in the late fifties, was the first Polish
professional manufacturer of computers. Next, Elwro
was the Polish biggest manufacturer of computers
(till late eighties). Therefore, the technology transfer
processes in Elwro must have been of major
technical and social aspects in the scale of all Poland.

The Authors consider that it may be worth wile to
follow the history of Elwro at the milestones defined
by launching of the most important products
(computers) manufactured by this enterprise. For
each milestone, the basic technical aspects of the
technology transfers have been discussed.

The final milestone was the fall down of Elwro. The
aspects of the technology transfer for this milestone
and some technical and social consequences of the
fact have been discussed.

2. THE PIONEERING PERIOD

During the pioneering period of Elwro (middle fifties
till late sixties), the technology transfer from the
developed countries to Elwro and to Poland was
severely limited. The reasons were of the political
and economical character. On one hand, there were
rather severe restrictions of embargo type which
resulted in limitations put on the technology transfer.
On the other hand, the negative foreign exchange
balance resulted in severe local restrictions on
importation from the developed countries. It may be
summarised that a zloty to dollar exchange rate was
assumed virtually as infinity and any home solutions
were recognised cheaper than imported ones.
Therefore, in practice, only the most necessary
electronic devices unavailable in Poland were
imported for Elwro.On the other hand, the transfer of
IT ideas was rather severely limited because of
political reasons (let us remind that these were times
of the cold war, Berlin wall, etc.). This, plus the
deficiency of convertible currencies, resulted in that
the design staff of Elwro could learn only the very
basic information on the IT available in the public
domain of the well developed countries.

In spite of that, the enterprise grew surprisingly well, launching several computer types and educating a big staff of electronic and computer engineers.

2.1 Elwro computers of the pioneering period

UMC-1 (Elwro, 1958). It was the first Polish industry computer developed in the electronic valve technology.

Odra 1001 (Elwro, 1963). The first Polish semiconductor technology computer using a drum as the working memory. Odra 1001 computers were rather successfully applied for numerical data processing.

Odra 1003 (Elwro, 1965). An improved version of Odra 1001, making use of a ferrite core memory. Successfully applied for numerical data processing.

Odra 1204 (Elwro, 1970). The first Polish computer of the up-to-date architecture, interrupt-driven. Though its hardware was realised in the discrete germanium device technology, the computer was very reliable and durable (some units worked satisfactorily for ca. 20 years). It found its application in professional numeric data processing (the Odra 1204 Algol compiler was recognised to be one of the best implementations for this programming language of that time in the world), some business processing application and even in pioneering industrial process automation (Wojsznis and Lewoc, 1972). For the computer, the first Polish operating system was developed.

The logical solutions applied in Odra 1204 were, in many cases, better than those used in the most popular minicomputer of that time (PDP 11). For instance, the interrupt handling system was simpler and, consequently, faster than that in PDP 11. Odra 1204 defines the milestone finishing the pioneering period of Elwro and, at the same time, of computers in Poland.

2.2 Basic technical and social aspects of the pioneering period

The severely limited IT transfer of that time was, surprisingly, of some major positive technical and social aspects on the overall scale of Poland. To develop computers, Elwro had to solve self-dependently many major technical problems, starting from development of logic modules needed to built sophisticated logical structures, printed-circuit boards, mechanical solutions of computer cabinets, interconnection wiring, memory systems, peripherals, up to warrant and post-warranty service. Therefore, the enterprise had to educate its own hardware design

staff and soon became the leading electronic design centre in Poland. This resulted in that young electronic and computer science graduates of that time who wanted to acquire profound know-how and experience in the domain of electronic system design and development used to choose Elwro as their first employer. After a few years they became efficient hardware designers and could be employed by other enterprises to solve complex electronic design problems. Thus, Elwro, which employed ca. 9,000 people at the end of the pioneering period, was the most important source of experienced electronic system design staff for the whole country.

Another technical and social aspect was that, in spite of the restrictions put on the IT transfer, there was a serious growth in computer applications in Poland. Because home made computer hardware was available, prospective users could develop solutions for their computational problems at reasonable and not "infinite" prices implied by hardly available convertible currency.

A major negative aspect of the IT transfer and of the most severe Elwro executive staff mistake were the limitations of computer applications due to lack of duly advanced software tools and applications. Elwro did not develop complex software tools nor applications. The burden of development of such software was on the customers. Some of them (in particular, the Wrocław University) could create software needed by many other customers (e.g. the Algol compiler mentioned above) but it took years before software needed for many applications appeared in the market. This resulted in higher costs of development of IT applications or even abandoning execution of IT projects.

3. ODRA 1300 FAMILY PERIOD

In early seventies, when the relations between Poland and the developed countries improved at a major extent, it became possible to increase the IT transfer. Elwro entered into co-operation with the International Computers Limited Company. Because the enterprise had to its disposal a numerous staff of hardware designers, it was possible to develop rather fast the Odra 1300 family of computers (Elwro, 1975) compatible with ICL 1900 Series computers. The hardware was designed by Polish engineers on the basis of logical specifications disclosed by ICL. The English Partner delivered a lot of software tools and application software, including business data processing software severely needed in Poland.

The decision undertaken proved to be a reasonable one and had many positive technical and social aspects.

First of all, Elwro had to maintain its top-level hardware design staff and, consequently, continued to be the main source of experienced electronic system designers in Poland.

Secondly, the Odra 1300 family computers, provided with ICL-made software, enabled rapid growth of numerical and, much more important, business computing in Poland (and in other Comecon countries). The computers became handy tools enabling many customers to solve their every-day calculating problems. Due to the availability of many application tools, development of individual systems became much less labour consuming and less expensive.

Another important and positive aspect of IT transfer, achieved in this case, was that Elwro and Polish customers did not limit their activities at distribution of OEM solutions only but also developed new solutions. A good example for an IFAC conference is that of the automatic control applications. Odra 1325 (ICL 1902a compatible), though initially intended for low-scale business and numerical processing only, was successfully adapted to industrial process control by Elwro who developed the real-time operating system EX2P and, in co-operation with the Computer Automation System Institute, process-control peripherals. Outstanding results were obtained by the Institute of Power System Automation in the Polish power industry (Lewoc, Rozent and Saczuk, 1992): for 15-18 years, Odra 1325 computers were operated in Polish Regional Power Distributions Centres, monitoring more than 50% power delivered by the Polish electric power network. The economic benefits obtained reached some US$ 40 billion (at the current price level) which is an amount considerable on the scale of the whole country.

A specific characteristic of the first two periods was that very good design and implementation staff must have been educated in Elwro and in Poland. Polish designers had much more difficult tasks to do in order to develop and implement successfully any novel system in comparison to their colleagues of well developed countries. So they had to learn more and could become better experts than their Western counterparts. This opinion was presented by Błach (1991) in Krems and was supported by several world leading designers of CIMs.

To summarise, the basic positive aspects of the IT transfer in this period were:
- rapid growth of computing science applications in business and science,
- continued training of experienced hardware staff for the Polish economy,
- economic benefits due to low-cost application software tools,

- new computer application domains (e.g. automatic control) and consequent economic benefits.

4. THE UNIFIED SYSTEM PERIOD

In late seventies, a political decision was undertaken to implement the Riad series computers (IBM /360 and /370 compatibles) in Comecon countries. Due to this, Elwro was to abandon manufacturing of the Odra 1300 family computers and develop (on the basis of logic specifications), instead, Riad computers (R-32 and R-34, see Elwro (1982)). The decision was fatal and had serious negative aspects.

The primary reason for that was the very poor quality of the operating systems OS MVT and OS VS. The operating systems included many poor, "baroque" and even erroneous solutions which decreased the performance characteristics of Riad based systems, which, even in theory, could not be too good because of not-necessarily redundant, time-ineffective hardware solutions. A spectacular evidence for that may be the Phoenix system developed by the Cambridge University: due to a series of modifications of the operating system and the access method, some performance characteristics in the interrogating operating mode were improved by several dozen times! (Lewoc, 1984).

In Poland, this resulted in high increase of computing costs, especially for on-line processing. This could be proved even by simple comparative performance evaluation experiments on Odra Family computers and on Riad computers (Lewoc, 1990): the costs of on-line Fortran and Pascal program development and debugging on R-34 were at least 10 times higher than on Odra 1305.

Thus, the technical and social aspects of the IT transfer depend, obviously, on that if the technology is good or poor. In the case under discussion, the decision on the technology to be transferred to Poland was undertaken by politicians who had neither knowledge nor experience necessary to evaluate the technology. They based their opinion on the magnitude of the technology creator and made a severe mistake of major negative impacts on the economy of Poland and other Comecon countries.

The basic technical and social aspects of technology transfer during that period were:

Loss of confidence into Riad computers. Expensive hardware which tripped rather frequently plus sluggish software made many users to stop believing in profitable solving of their computational problems on Riad computers.

Serial economical losses. Since the decision on use of Riad computers was a political one undertaken on the top hierarchy level, governmental computing centres had to buy this hardware. The decreased customers' confidence and much higher data processing costs than those on Odra 1300 family computers resulted in serious decrease in the demand for computing power in Poland. This resulted in that the computer systems of poor performance utilised at a small degree, which resulted in rather serious economic losses in Poland and other Comecon countries.

Continued education of experienced hardware designers. Since the hardware of Riad computers was more complex that that of Odra Family computers, the demand for top-quality hardware design staff was even higher than before. Due to this, Elwro had to educate experienced hardware designers and continued to be a source of experienced electronic system designers and service engineers for the whole country. However, due to the lack of success with the Riad Family computers, the experience acquired by electronic systems designers and service men in Elwro ceased to be a pass opening doors to any job in electronic systems in Poland.

5. DECADENCE PERIOD

5.1 The falling down process

The negative impacts of launching of the Riad computers in Poland resulted in a rather poor technical and economical condition of Elwro at late eighties. This condition was made worse by other factors of that time.

The most important thing was the beginning of the Personal Computer are in the world. The market for main frame computers decreased around the world. The decrease was even bigger in Poland since the available main frames were poor ones. At the same time, the customs duty for PC-s was cancelled and companies and individuals in Poland could easily import personal computers. Elwro could not manage to enter the segment of PC manufacturing and was pushed off the computer hardware market.

A major factor of economical and political character was the naïve liberalism of the government at that time. All import restrictions in the IT domain were released, the dollar exchange rate was established at some finite level and Elwro had to compete alone with all computer hardware manufacturers of the world.

One more economic premise of that time was the monetary policy of the government. Earlier operation of Elwro, as in the case of other Polish enterprises, was based on low-cost credits. When credits became very expensive, Elwro had no financing source neither for research and development nor for even current operating activities. The government-owned enterprise which got no support from the government had to fall down.

This last process was as follows:

Northern Telecom was interested in buying the residual Elwro staff (ca. 1,500 employees) and premises, to establish there a local manufacturing and sales centre for NT digital telephone exchanges and initiated technical and managerial procedures oriented towards such merge. However, the government decided to sell Elwro to Siemens (1993) and to grant to the company the permissions for sale of telephone exchanges in Poland.

In spite of the investment obligations included in the Elwro sales contract, Siemens never started any new IT activities in ex-Elwro. On the contrary, to bypass the employment guaranties, the company started soon (1993) the voluntary job quitting program on the basis of rather high severance pays. Shortly, ex-Elwro employed only 250 peoples involved in mechanical and chemical type of work (manufacturing of cases for electronic equipment). This residual staff and the premises were finally sold to Telect.

At that time, ex-Elwro people produced rather a black joke: A part of Elwro land was earlier possessed by the municipal cemetery. When this are was regained by the municipal authorities, ex-Elwro people started to claim that it is the only place where they can meet. Which was actually the fact since the authorities treated employment in Elwro as a sufficient condition for allocation of the last allotment there.

5.2. Major technical and social aspects

We can not deny that personal computers made a revolution in the domain of the Information Technology. However, the aspect of destroying the Polish biggest manufacturer of computers during the IT transfer was a very harmful one for the overall economy of our country and actually not necessary as evidenced by the experience of other ex-Comecon countries.

The perspective manufacturer of main frame computers in Poland disappeared. This was a major mistake: after the very initial period of rapid growth of personal computer applications, the users realised that, for large-scale computing applications, personal computers are much more expensive and complex, integrated computing systems are hardly feasible.

Another very negative aspect of the IT transfer in this period and later was that there happened some very harmful changes in the approach to development of IT systems in Poland. The Polish designers with their aim, know-how and aptitude to develop their own systems were substituted by a lot of young, rather incompetent people specialised in earning money on distribution of foreign, if possible – stolen (pirating) solutions. Usually, the biggest computing science tasks in which such people are involved is the task of localisation, i.ę. the task for translators rather then computer people. A clear evidence for this change in approach to development of computer systems after liquidation of Elwro is that the present biggest manufacturer of PC hardware in Poland is, in fact, only a distributor of Computer Integrated Management software delivered by some European Software Development Company. This distributor even can not realise nor believe that there are local and world-wide problems which can be solved autonomously, using the experience of the staff educated on projects executed by Elwro and co-operating Polish companies.

The technical and social aspects of the facts are that Poland is now an economicly dependent country in the area of Information Technology. And the experience of our best experts in the computing science is being spoiled. For instance, we recognise that at least of 10 years of work on complex projects is necessary to educate an efficient leading designer of the operating systems. One of such Elwro experts prepares programs for computation of duty taxes, the other one – for computation of the Wrocław municipal communication utility; surely, they do useful work, but would it not be more beneficiary for our country if they developed operating systems for new IT applications?

An economic aspect of such IT transfer are the severe losses suffered by the Polish economy in development of novel computing science applications. They are prepared for foreign applications first and adapted / customised to the Polish conditions later. Therefore, we have to wait for novel solutions to be verified in foreign countries and then adapt them to the Polish requirements. In any case such process is more expensive and, possibly, its results are poorer for the Polish users than in the case that the systems are designed and developed directly for local applications.

CONCLUSION

The technology transfer process in IT may have very harmful aspects for overall countries. Therefore, it should be monitored and controlled by the governments involved.

In particular:
- local staff should be involved at a high degree in reception, development and implementation of the new technologies,
- in the era of serious technology transfer processes, local governments should support their own manufacturing and development industries to ensure them a fair chance in competition with large-scale, powerful foreign manufacturers,
- the fact that local manufacturers' managerial staff may be rather incompetent due to the monopolist position of the manufacturers earlier doe not relieve the owners (governments) from the responsibility for the enterprises; on the contrary, in the difficult circumstances of the technology transfer process, the owners should monitor the activity of the managerial staff and, where necessary, change the staff to more competent and skilful one,
- neglecting to support local manufacturing and development enterprises in the era of the technology transfer will, almost for sure, result in high economic losses of the country involved, through spoiling of the high-cost experience of local experts, delays in distribution of novel solutions, rather poor match to local conditions and, even worse, in economic dependence of the technology receiving country.

REFERENCES

Błach, L.K. *et al.*, (1991), West-East approaches of automation, A case study, *CAA (IFAC)*, Krems,

Elwro (1958), UMC-1 computer, *Operating, Maintenance and User Manuals*, Wrocław.

Elwro (1963), Odra 1001 computer, *Operating, Maintenance and User Manuals*, Wrocław.

Elwro (1965), Odra 1003 computer, *Operating, Maintenance and User Manuals*, Wrocław.

Elwro (1970), Odra 1204 computer, *Operating, Maintenance and User Manuals*, Wrocław.

Elwro (1975), Odra 1300 computers, *Operating, Maintenance and User Manuals*, Wrocław.

Lewoc, J.B. (1884), Quo vadis Jednolity Systemie, *Informatyka No. 3*, Warszawa.

Lewoc, J.B., (1990), Performance evaluation problems for actual computer networks, *AMSE Press*, Tassin.

Lewoc J,B., M. Rozent and I. Saczuk (1992), The computing power and the computing systems for the power industry in Poland, *MICC (IFAC)*, Prague.

Wojsznis, W., J.B. Lewoc (1972), Sistema upravlenia prodvizeniem metalla v prokatnom cehe, *IV Int. Conf. of Comecon Countries on Process Automation in Black Metallurgy*, Zaporoze.

IFAC
▷○▷
Publications
www.elsevier.com/locate/ifac

TECHNOLOGY CHANGE, TECHNOLOGY TRANSFER AND ETHICS:
PART I

M.A. Hersh

*Centre of Systems and Control and
Department of Electronics and Electrical Engineering,
University of Glasgow, Glasgow G12 8LT, Scotland.
Tel: +44 141 330 4906. Fax: +44 141 330 6004. Email: m.hersh@elec.gla.ac.uk*

Abstract: This is the first of a two part paper on technology transfer and technological innovation. The first part considers issues relating to the role of technology in society, the interrelationship between political and technological change, values biases and the role of technological and scientific expertise. The second part will consider environmental and social impacts of technology change and transfer. *Copyright © 2001 IFAC*

Keywords: Technology transfer, innovation, ethics, policy, consultation.

1. INTRODUCTION

Over the past few decades there have been dramatic developments in available technologies and human activity has put considerable stress on the natural ecosystems of the planet. It is therefore useful to consider the role of technology in society at present and what this could and should be in the future. This includes issues of accountability and transparency in terms of who makes decisions about technology policy, the responsibility of such decision makers to the wider population, how the population as a whole influences this process and the availability of clear accessible information about specific technologies and their advantages, disadvantages and limitations. A very important aspect of this role is the impact of changes in technology on social structure, culture, attitudes and power dynamics, as well as the different political, social and economic forces that decide which types of technological change will be encouraged and supported.

A related issue is education about technology and how this can be used to empower people to participate in the decision making and policy

formulation processes. Knowledge transfer is often seen as a one-way process from 'experts' to the ignorant lay population or from experts in the 'developed' countries to both the lay population and scientists and engineers in 'developing' countries. However in order to be successful knowledge transfer should be a two-way process which takes account of and communicates the knowledge, experience and expertise of both sides. The above discussion has illustrated some of the tradeoffs and different interests involved in technology change and technology transfer, making it essential that they should be considered in an ethical framework.

The first part of this paper will focus on the role of technology, technological change and technology transfer. Part 2 will consider environmental and social impacts of technology.

2. THE ROLE OF TECHNOLOGY IN SOCIETY

Technology is having an increasing impact on human society and the natural environment. It can be used to promote particular values and interests, implement and reinforce or reduce discrimination and increase or reduce inequalities. For instance a particular New York bridge was built too low for buses, so only those with cars could get to the park beyond it (Winner, 1985), presumably to keep out poor people. Most buildings are not designed to be energy efficient and are generally not designed for wheelchair access and many existing technologies and devices have not been designed to be used by people with disabilities. Technology can affect life chances by increasing or removing opportunities and increasing or reducing risks (Street, 1992). Dependence on technology can lead to inequality, as those who control technology have power over its users and those without access to a particular technology are disadvantaged through exclusion.

There are a number of different theories of the relationship between political and technical change (Street, 1992). On the one hand theories of autonomous technology consider that technology has its own rationality and acquires an independent momentum which allows it to order all human activity, including politics, and act as the driving force in social choices. On the other there are theories of political choice in which political decisions are the driving force which determine the nature of technology. However the relationship between politics and technology generally has elements of both these theories, with complex feedback loops, which also take account of other factors such as culture and social organisation, linking technological and political changes.

Thus, political and other choices can have a significant effect on determining technology change, but existing technology generally contributes to determining the political and social context in which decisions are made. Many governments and political parties have technology policies, as technology is now considered too important to be left to technologists or industry (Braun, 1995). The prioritisation of ethical considerations, including determining and prioritising real needs, reducing inequalities and protecting the environment is unfortunately rarely the most conspicuous element of technology policies. The development of technology is generally facilitated by and often totally dependent on the availability of funding and other resources. Views about whether government or industry should fund applied research are generally dependent on political ideology and views of the type of society that is desirable. However source(s) of funding often have a significant effect on research directions and strategies and increasing commercial and military sponsorship is affecting the independence of research and often the ability to publish freely, and may even lead to a preference for certain types of results. If appropriately targeted (which is relatively unlikely in the present socio-political and economic climate), government funded support for research could be

used to focus on the solution of urgent social and environmental problems.

However many real social and environmental problems, such as famine and malnutrition require the political will to make changes and more equal access to existing technologies, rather than the development of new technologies. In other cases, such as HIV and AIDS, there is a lack of resources and economic interests, such as the large pharmaceutical companies, have ,sometimes unsuccessfully, tried to bar access to affordable treatments through patent restrictions. There is also a need for the recognition of indigenous technologies, such as traditional irrigation techniques in Kenya (Ilkkaracan et al, 1995) and the need for relatively small amounts of funding to implement them.

Even in the absence of specific policies, choices about research directions and technological change are influenced by the socio-economic and political context. In some cases technology development may seem to be autonomous due to a lack of control by the political system, (McNeill, 1983). The complex interactions between politics and technology mean that new technologies may both be developed in response to social change and themselves cause social change (Hughes, 1987). Another important factor is inertia, which will result in a tendency for technological and other developments to continue in existing well established directions and resist change. There is generally a tendency for previous technology choices to inform future choices and pressures to continue with existing technologies, so as not to waste the resources already expended, regardless of whether they give the best option. There may also be pressures to use a particular technology even when it is not particularly appropriate to do so, so as to justify the associated capital costs, particularly if they are high. This happens at both the organisational and individual level and in the case of inappropriate use of medical technology may even increase risks to patients (Hellerstein, 1987).

Some technologies may be neutral, with the impact and the nature of many of their effects dependent on how they are used. However, certain types of technology, such as weapons technologies, have been developed for particular applications and cannot be considered neutral. Since some applications of apparently 'neutral' technologies may not be benign, the individuals or organisations developing and implementing them have a responsibility to try and control the nature of the applications of these technologies, though they may not be in a position to do so fully. Although it is sometimes assumed that technological innovation is intended to meet human needs, in practice innovation has frequently resulted from competition and then been sold to a sometimes reluctant public, and has had little impact on real human needs (Braun, 1995).

There is a tendency for users to become dependent on new technologies and to experience problems if the technology is withdrawn. Particular examples include the television, washing machine and modern computing equipment. On the other hand appropriately designed technology can increase available opportunities, including to sections of the community who would otherwise be disadvantaged. For instance assistive technologies and accessible technology design can be used to extend the opportunities available to people with disabilities.

Dependency on particular technologies can lead to an imbalance of power which allows the suppliers of technology to apply different safety standards to the same technology in different countries. For instance, the safety standards in the Union Carbide factory in Bhopal were much less strict than those in the US (Martin et al, 1996) and medical technologies which have been refused licenses on safety grounds in the industrialised countries are often imported into 'developing' countries. Different safety standards are often applied to different technologies and are also affected by the context in which they are used. For instance oil platforms are designed to lower safety standards than would be permitted on land (Observer, 10 July 1988). This means that workers are likely to experience avoidable risks in the work place, the size of which depends on the type of technology and the country they are working in, raising issues of why the most stringent safety standards and protective measures cannot be transferred between countries and technologies. While there may be valid reasons for taking into account local factors and cultural differences in the interpretation and detailed implementation of safety regulations, it is clearly unethical that lesser safeguards should be applied in some countries or to some technologies.

Many governments, employers and professionals seem to have a preference for technology which allows or helps them to increase control over users, clients or employees. For instance the nature of modern medical technology can act to restrict a patient's choices and, for instance, in the case of childbirth, turn a natural process into something that is practically automated. This of course does not negate the importance of the appropriate use of technology in childbirth in reducing infant mortality. Government preferences and considerably higher research budgets for nuclear power as compared to renewable energy sources may be a consequence of the fact that the inherent risks of the technology allow workers to be vetted and union rights, such as the right to strike, to be removed. For instance the UK government used support for nuclear power as part of a strategy for reducing the influence of the

previously powerful National Union of Mineworkers. The other side of this is the fact that the potential dangers of a particular technology, such as nuclear power, may lead to restrictions on workplace organisation and the removal of the right to strike (Winner, 1985). However restrictions of this type on workers' organisation are an infringement of their civil liberties and raise questions as to whether it is ethical to use technology which requires such restrictions.

3. ACCOUNTABILITY, TRANSPARENCY AND DECISION MAKING

The importance of technology is not always reflected in the associated decision making processes, giving rise to issues of how best decisions about technology should be made and whether new types of decision making processes are required. Some of the most important issues relate to making decisions about technology to develop a type of society which meets the needs of all segments of the population in an equitable way and with minimal impacts on the natural environment. Technology projects often affect different groups of people differently and in many cases there may be considerable opposition to policies favoured by governments.

Most decision makers or politicians and members of the general public lack technical expertise. A number of spectacular and well publicised accidents or near accidents, including Bhopal, Three Mile Island and Chernobyl have contributed to unease about technology. There is often a tendency to dismiss public concerns about technology as misinformed or resulting from ignorance and a lack of understanding. This is clearly unethical. In addition the general public often seems to have a good understanding of risk and uncertainty (ESRC, 1999). However the issue still remains as to how best to take account of both technical expertise and the views of the lay population in decision making on technology issues. There is also a need for both short and long term technology planing, whereas most economic and political systems encourage short term thinking.

Governments frequently try to control science and technology, including by using financial constraints to put pressure on scientists to follow particular lines of research in order to obtain funding (Boehmer-Christiansen, 1992). Since different scientists and engineers often disagree, governments can generally find a body of scientific opinion which supports their policies, for instance on nuclear power. Lack of information or disagreement between scientists has often been used by governments as a reason not to take action, for instance, on global warming. Scientific reports have also been used to reassure the population about the effects of pollution incidents such as the Chernobyl disaster. Therefore independence of scientific and engineering judgements from political control is vital for public confidence. This, however, does not imply that political concerns should not affect decisions on research directions

Technology assessment arose out of the need to fill the information gap for policy makers and aims to provide an interdisciplinary analysis of the full range of consequences to be expected from a technology in different circumstances (Braun, 1995). However this analysis, though useful, is unlikely to be totally independent of value bias. Many 'experts' are influenced by political or ideological views or have pet theories which may effect their conclusions. Political, social and/or scientific biases may sometimes lead to the distortion or misinterpretation of results to support cherished scientific or ideological theories, not always through conscious manipulation, as in the case of experiments to demonstrate racial differences in brain size (Gould, 1984). Expertise is socially created, with what is considered authoritative knowledge largely determined by social values, and certain types of knowledge and experts generally receiving greater official acceptance than others. It should therefore be recognised that acknowledged 'experts' do not have a monopoly of knowledge. Increasing specialisation is making it more difficult for experts to value a technology as a whole, both due to frequently only having expertise in too narrow an area and threats to a particular technology sometimes putting careers at risk and therefore making specialists less independent.

Approaches to decision making can embody different degrees of public and expert participation and public participation may be exercised directly or by elected representatives, who may have their own perspectives and vested interests. The decisions reached are likely to be effected by human factors, existing paradigms and scientific orthodoxies and views of what is acceptable science (Rose et al, 1984) and sometimes distorted by the tendency to try and make sense of new evidence in terms of existing paradigms. Political and social values will also have a significant effect on the agenda in terms of the questions to be considered and problems to be solved (Street, 1992). However this does not mean that science and technology should be totally divorced from political concerns, particularly since choices of problems to investigate have considerable real world effects and the unwise use of technology has led to severe environmental and other problems. Rather there should be more openness about the relationship between science, technology and politics and any underlying political or ideological agendas.

4. TECHNOLOGY TRANSFER

Technology transfer can be defined as the multilateral flow of information and techniques across the boundaries of science, technology and the practical world and can involve either hardware or software transfer (Khadkikar et al, 1980). Successful technology transfer should involve the technology functioning as intended, mastery of the technology ('know how') and the ability to develop the technology and adapt it to new circumstances ('know why') (Wilhite et al, 1992). Decisions on technology transfer should also involve consideration of ethical issues relating to the distribution of benefits and costs and environmental issues, such as pollution, damage to micro-environments and energy efficiency.

Technology transfer can occur in a variety of different contexts. However it is often seen as a one way process from the powerful and knowledgeable to the powerless and/or ignorant. It is generally assumed that technology transfer from the industrialised to the 'developing' countries is in the interests of and brings benefits to the people of the 'developing' countries. However this depends on the types of technology transferred, how they are transferred and who they are transferred to. Not all technology is appropriate to local conditions. The long and short terms aims of technology transfer and their relationship to wider political and social goals should be clear and serious attempts made to avoid the mistakes of the industrialised countries in terms of the environmental and social costs of technology. However, the 'developing' countries can still benefit from transferring appropriate technologies from the industrialised countries (Meshkati, 1986).

In order for technology transfer to be successful the technologies transferred should be appropriate to local conditions and transferred to the people who need them. Considerable efforts may be required to ensure that the benefits of technology transfer are widely and equitably distributed. In many cases this will require the transfer of different types of technology. Transfer of high level technologies may have a significant impact on the overall economy and balance of payments, but little meaningful effect on the quality of life of the majority of the population. Improvements in general living conditions may require transfer of low level or intermediate technology at local level, often to groups of women. There may also be benefits in transferring technology from developing or newly industrialised countries, where conditions and working practices may be relatively similar, or from other parts of the same country. The importance of consulting and taking account of the needs and wishes of people at the local level, rather than decisions being made solely by government or senior management of the organisations making the transfer should also be stressed.

The introduction of new technology, particularly into a small rural community or over a short time period, can have significant negative impacts and may lead to considerable disruption of lifestyles and threats to local culture, depending on how the transfer is managed. This possible disruption should be considered when the effects of the technology are evaluated, as well as impacts on the local economy. Expectations of the technology and the type of society it is intended to bring about should be clarified before the technology is introduced, but this does not always happen.

Using local workers can contribute to developing skills, providing employment and avoiding dependency. However it may be necessary to also bring in (skilled) workers from outside. Particularly in small, relatively homogenous and/or isolated communities this can have a significant negative or even disastrous impact on the local economy and community (Eschenback et al, 1988). The organisations carrying out the transfer have a responsibility to minimise any negative impacts. This responsibility should include the provision of training for local people. The pace of change is likely to be significant with gradual incremental changes being less disruptive than dramatic changes over a short time period. However the nature of a particular technology will often determine whether it is possible to introduce it in stages. This will be difficult for some technologies, such as oil pipe lines.

Ethical considerations need to be taken into account when making decisions on technology transfer. In particular the balance of benefits and costs between different population groups should be considered. It should be noted that principles of fairness imply that one group of people should not be exposed to the risks and another receive the benefits, however significant these benefits may be (Rawls, 1971). Therefore importing technology into an area which results largely in benefits to those outside it is generally not justified in ethical terms. The balance of benefits within an area should also be considered and it not automatically assumed that 'modernisation' will be overall of benefit. There are also issues of the pace of change. In general relatively small scale managed change is likely to bring more benefits and have less disadvantages than the introduction of technology over a fast time scale.

Ensuring that the technology functions as intended is often much easier that transferring know-how and know-why, due to a mixture of differences in the donor and receiver contexts (Wilhite et al, 1992) and cultures, as well as details of the transfer and sometimes opposition from the donor. Donors may have concerns about transferring full know-how and know-why, as this will transfer control of the technology to the recipient and may create competitors. There may also be preferences for transferring a slightly older version of the technology or not transferring the most recent developments (Wilhite et al, 1992) for similar reasons. However this raises ethical issues, as it would contribute to perpetuating the technology gap and possibly also to maintaining dependence, as the recipient country may be in a position of having to apply to the donor for updates. Responsible donors should inform recipients of the likelihood of a particular technology becoming obsolete or being replaced in the short term, as well as trying to support them in making the best technology choices.

All organisations operate within a wider cultural and national context, which will generally affect attitudes, working practices, personal interactions and what is considered 'normal' and/or acceptable behaviour. There may also be cultural differences in organisational structure, decision making, attitudes to and the organisation of work (Meshkati, 1986) between organisations in different cultures and nations. In many countries the working year is broken by public holidays, often based on religious festivals and/or fasts, which will be different in different cultures. This can lead to problems in transfer between cultures with different religious and other festivals. Differences in research culture and the relationship between science and technology in different countries can also impede transfer (Wilhite et al, 1992). There is often considerable ignorance in the industrialised countries about life in the 'developing' countries, supplemented by stereotypes.

Therefore in depth education about local life styles and conditions should be a prelude to technology transfer. This should be followed by the development of communication strategies across cultures and a recognition that different cultural styles are equally valid. Any tendencies for donor organisations to try and encourage or impose their working, religious and/or cultural practices on recipients should be resisted.

People in many developing countries are accustomed to craft-oriented industry and do not work particularly efficiently on modern automated machines, which they feel degrade them from craftsperson to labourer (Baranson, 1969) or may dislike factory type jobs which force them to put in time rather than work to complete a task (Meshkati, 1986). This raises issues about the nature of work and the need for working practices which show respect for workers. It highlights questions about the effects of modern automation technology on working conditions and shows the importance of technology transfer being a two or multi-way process. Developing strategies to adapt traditional craft based working practices to modern automated machinery could be of benefit to workers in the industrialised countries.

61

New technologies generally have environmental and social impacts, so that technologies with minimal negative impacts should be chosen and implemented in a way that further minimises these impacts. Recipient and donor organisations could be considered to share responsibility for this. However donor organisations should take responsibility for transferring information on environmental issues associated with the technology, such as waste and emissions reduction or control strategies and energy reduction. This is particularly important, as industrialised countries frequently have greater expertise in these areas than 'developing' countries and could stimulate cooperation on technologies with minimum environmental impact (Wilhite et al, 1992).

There has been a tendency for technology to be designed for a norm, with regards to physical attributes and other factors, which is probably only valid for a minority of the world's population. A number of studies have demonstrated the physiological differences between different population or ethnic groups (Sen, 1982; Pierce, 1963). Most technology is designed for men and consequently unsuitable or not best suited to women and does not take account of the needs of people with disabilities. It is important that technology design takes account of such differences in a value neutral way that does not consider certain body types or sets of physical abilities superior to others. Otherwise existing inequalities in access to technology will be strengthened, since, for instance, American anthropometric design standards are suitable for only 10% of the largest Vietnamese (Wilhite et al, 1992) and presumably hardly any Vietnamese women. There are also issues of design for different and sometimes extreme environments (Eschenbach et al, 1988) and the fact that design principles that work well in one environment may not adapt easily to a different one, particularly if the physical conditions are very different.

CONCLUSION

The relationship between politics and technology is based on complex feedback loops. This implies that society can significantly effect the nature and extent of technology change, but this change is partly determined by existing technologies. A dependence on particular technologies may lead to an imbalance of power, particularly when there is a preference for technologies which facilitate an increase of control over users, clients or employees. Unfortunately the general public and most decision makers lack technical expertise. However the general public has a good understanding of the issues of risk and uncertainty, which are important in decision making on technology. There is also considerable suspicion of technology and governments frequently try to control science and technology. For technology

transfer to be successful, the technologies involved should be appropriate to local conditions, take account of local cultural and other values and be transferred to people who need them. Consultation at local level is also important, as is the recognition that technology transfer should be a two way process, with learning in both directions.

REFERENCES

Baranson, J. (1969). *Industrial Technologies for Developing Economies*, Praeger Publishers.

Boehmer-Christiansen, S. (1992). *How much 'science' does environmental performance really need?* SPRU, University of Sussex.

Braun, E. (1995). *Futile Progress, Technology's Empty Promise*. Earthscan Publications.

ESRC (1999). *The Politics of GM Food*, Special briefing no 5, www.gecko.ac.uk

Eschenbach, T.G., G.A. Geistauts and G.N. Jones (1988). Technology transfer in Alaska, *Tech. Management Publication*, **1**, 336-344.

Gould, S.J. (1984). *Mismeasure of Man*, Penguin.

Hellerstein, D. (1987). Overdosing on medical tech. In: *Contemporary Moral Controversies in Technology*, (A.P. Iannone, ed.) OUP.

Hughes T. P. (1987). The evolution of large tech. sys. In: *The Social Construction of Tech. Systems*, (W.J. Bijker et al., eds.), pp. 51-82. MIT Press.

Ilkkaracan, I. and H. Appleton (1995). *Women's Roles in Technological Innovation*, Intermediate Technology Publication.

Khadkikar, S. and D. Mukhedkar (1989). Canadian examples of international transfer of technology. *IEEE Eng. Management Conf. Record*, 22-25.

McNeill, W.H. (1983). *The Pursuit of Power*. Basil Blackwell.

Martin, M.W. and R. Schinzinger (1996). *Ethics in Engineering* (3rd ed.), McGraw-Hill.

Meshkati, N. (1986). Major human factors considerations in tech. transfer. *Human factors in Org. Design and Management* **II**, 351-367.

Pierce, B.F. (1963). The ethnic factor in the man-machine relationship. General. Dynamic Corporation, San Diego, California

Rawls, J. (1971). *A Theory of Justice*, pp. 179-183. Harvard University Press

Rose, S. L. Kamin and R. Lewontin (1984). *Not In Our Genes*, Penguin.

Sen, R.N. (1982). Certain ergonomic principles in the design of factories in hot climates. *Int. Symp. on Occ. Safety, Health & Work. Cond.*, ILO, Geneva.

Street, J. (1992). *Politics & Technology*, Macmillan.

Wilhite, H. and A. Stenseth (1992). Barriers and policy issues ... transfer of advanced energy tech., *Electronics Info & Plan.*, pp. 681-699.

Winner, L. (1985). Do artefacts have politics? In: L D. MacKenzie et al. (eds). *The social shaping of technology*, pp. 26-38. Open University Press.

IFAC

Publications
www.elsevier.com/locate/ifac

COMPUTERIZATION OF METALLURGICAL PRODUCTION AS THE MEANS FOR SOLUTION OF THE ENERGY-SAVING AND ECOLOGICAL SAFETY TASKS

S.A. Vlasov, A.L. Genkin, and N.G. Volochyok

Institute of Control Sciences, Moscow, Russia

Abstract: Energy-saving and ecological safety of the metallurgical processes have been in the focus of the ever growing attention recently. The use of computer and information technologies for solving these problems is logical for two reasons. First, the modern computers and the computerized control systems using them can solve these problems if they are specified as criteria of the functioning of these systems. Second, various new technological processes, facilities, systems and complexes contributing to the solution of the energy-saving and ecological safety problems are developed and start already being marketed, and yet their use is not conceivable without modern computer and information technologies. The paper is concerned with example of using computer simulation for decision of problem of energy-saving in "steel-rolling mill" complexes. *Copyright © 2001 IFAC*

Keywords: computer simulation; energy-saving, steel-rolling complexes, control efforts, control algorithms.

The world experience shows that about 40% of the control tools and systems at metallurgical plants already meet the manufacturing safety, energy- and resource-saving standards. Energy-saving and ecological safety of the metallurgical processes have been in the focus of the ever growing attention recently. The use of computer and information technologies for solving these problems is logical for two reasons. First, the modern computers and the computerized control systems using them can solve these problems if they are specified as criteria of the functioning of these systems. Second, various new technological processes, facilities, systems and complexes contributing to the solution of the energy-saving and ecological safety problems are developed and start already being marketed, and yet their use is not conceivable without modern computer and information technologies. The examples include ROMELT, CSP-plant processes, twin-shell electric furnace, compact blast furnace etc. A notion of the energy-ecological quality of a process has been coined and its improvement has become the aspiration of the designers of technological

processes, hardware, and control systems, and the managers of different levels responsible for business processes of individual plants.

On the other hand, the rapidly growing cost of energy resources all over the world makes the energy-saving one of the critical components contributing to the competitiveness of metallurgical plants.

Energy-saving is a complex problem with social, economic, technological, ecological and other aspects to it.

During a crisis the role of the production control and organization grows significantly. With the stable economy, the energy-saving effect of the improved control is estimated by different experts as equal to 5-10 to 30%; in Russia today about 75% of the excessive energy use in metal production is caused by the production control problems.

The experience of the Institute of Control Sciences of the Russian Academy of Sciences shows that the computer simulation of specific features of technological processes allows developing efficient control systems taking care, among other things, of energy- and resource-saving.

Designers of modern metallurgical complexes and computerized control systems for them can also be offered a simulation system described in (Vlasov and Belov, 1995). This system uses the simulation software developed and constantly updated by the Institute of Control Sciences. This simulation system covers metal production from steel melting and ladling to strip and section rolling mills (for different technological schemes and different hardware combinations) and, hence, can be used to estimate the energy-saving and ecological safety efficiency of different approaches.

An example is offered by the Control System Used for Energy-Saving Technology (abbreviated in English as SUET) for hot strip rolling mills using slabs fed by the heating furnace. This paper discusses the main principles and algorithms of this system.

Assuming a strip rolling complex to be an object of control of organizational-technological systems we can distinguish main sections of metal processing in the technological line "slab reheating - hot strip rolling": 1) reheating furnaces; 2) roughing train; 3) intermediate table; 4) finishing train.

As metal moves along this line the energy consumption in its different sections varies. In metal reheating prior to rolling the fuel consumption depends mainly on the slab parameters and slab production rate at the furnace output. The rolling of a workpiece in the roughing and finishing train is accompanied with essential electric energy consumption due to metal deformation. Thermal metal losses in the inter-stand gaps and on the intermediate table are due to the radiation and convection and depend on the rolling speed and workpiece thickness and also on parameters of equipment and control system furnaces-mill complexes (see Fig. 1).

Each of the sections has a respective control system. The aim of designing an energy-saving system for controlling the temperature-deformation mode in the hot wide strip rolling mill is to adjust the individual subsystems in accordance with the chosen criterion of optimality and coordinate their functioning depending on a specific production situation.

All conventional methods of reducing thermal losses in hot strip rolling can be classified into two types:

using additional equipment to reduce thermal losses;

optimizing the temperature and deformation mode of rolling which permits controlling the metal temperature.

In the latter approach the energy is redistributed between individual sections of the "slab reheating - hot strip rolling" technological line, while the technological process is optimized to meet the required quality of the finished product and power parameters of the mill. One of the most efficient methods in this approach is to vary the reduction mode in the mill stands (Genkin, et al., 1992).

The analysis performed has shown that methods used in both approaches are almost equally efficient, however, capital expenditures in the second approach are significantly lower, hence, it is preferable.

The optimality criterion for the initial mill adjustment has to meet various requirements to metal processing in different sections of the line. Taking into account the specific features of the rolling process the optimization criterion is chosen as a function determining the economic state of the production observing the quality requirements to the hot rolled product (geometrical sizes and physical-mechanical metal properties). The restrictions also include extreme values of the technological process parameters, mill design characteristics, and mill arrangement. Assuming the multi-criteria nature of the problem its solution can be approached mainly by heuristic methods. The main optimality criteria in rolling control in different sections of the technological line are as follows:

1. Maximal efficiency of the mill. This criterion can be met by strict coordination of the throughput of various sections of the technological line (furnaces, roughing and finishing trains).
2. Minimal fuel consumption in the furnaces. This is required to guarantee an optimal metal heating mode at a given slab temperature.
3. Minimal power consumption in the main drives of the rolling stands. This criterion can be used when it is not possible to change the metal heating temperature or in some other situations.
4. Minimal total fuel and power consumption for metal heating and rolling. This criterion includes fuel consumption for the slab reheating in the furnaces prior to rolling, and power consumption in the main drives of the rolling stands taking into account additional metal losses in its heating.

In some situations there may be a need to use the criteria of uniform stand loading, rolling at a fixed temperature at the input and output of the trains, minimize heat losses, etc.

To optimize an initial state of a technological line "metal reheating - hot strip rolling" the author used mathematical models to simulate the metal heating and rolling, as well as the control system functioning. The domain of acceptable controls is determined by the rolling forces and torques, capacity of the main drive of the stands, the roll capture conditions and the length of the inter-stand tables. The variation of the reduction mode in rolling stands is taken to be the control input. A heuristic procedure of the reduction variation in rolling stands is developed which provides for the redistribution of the stand loading in accordance with the chosen optimality criterion.

The following modes of the initial adjustment of the stands can be used in the process of metal processing in a strip rolling "furnaces - mill" complex:

1. Rolling with fixed reductions. It can be implemented in the roughing and finishing trains in accordance with the technological instructions for metal heating and rolling.

2. Energy-saving mode. It can be used in roughing and finishing stands to minimize total fuel and power consumption for the slab reheating and metal rolling.

3. Fuel-saving mode. It can be used in roughing and finishing stands to minimize thermal losses. It minimizes the fuel consumption and metal scaling in reheating furnaces.

4. Electric energy-saving mode. It can be used in roughing and finishing stands to minimize electric energy consumption in main drives of rolling stands.

5. Full loading mode. It can be used for full loading of the stands by one of the parameters. It is designed first of all for the roughing train, since the finishing train cannot be fully loaded due to technological reasons.

6. Uniform loading mode. It can be used in roughing and finishing stands for the uniform loading of all stands of the train by one of the parameters.

7. Controlled rolling mode. It can be used in roughing and finishing stands to support meeting the specified metal temperature at the input and output of the respective train.

In view of the above and since the total metal heating and rolling expenditures are mainly caused by the variation of the reduction mode in the roughing train rather than in the finishing one, it is obvious that the roughing train is more representative from the point of view of selecting the initial adjustment modes of the stands.

The Fig. 2 illustrates example of results of SUET application.

REFERENCES

Vlasov S.A. and A.D. Belov (1995). Computer simulation in CAD-CAM-CAPP systems for steelmaking. In: *Preprints of the 8-th IFAC Symposium on Automation in Mining, Mineral and Metal Processing*. Sun City, South Africa, pp. 139-144.

Genkin A.L., A.R. Kudelin, O.D L'vova., *et al.* (1992). Efficiency of energy-saving technology in hot strip rolling. In: *Preprints of the 7-th IFAC Symposium on Automation in Mining, Mineral and Metal Processing*. Beijing, China, pp. 236 – 238.

EQUIPMENT   CONTROL SYSTEM

Optimization of the range of products

Optimization of the equipment composition used in the CTC "furnaces-mill"

Optimization of the coordinating parameters for the CTC "furnaces-mill" (total length of the effective hearth of the furnaces, slab temperature and thickness, the thickness of semi-finished products)

FURNACES

MILL

Choice of the furnace type and optimization of the furnace parameters

Choice of individual parameters of the mill equipment

Choice of optimal heating modes

Stabilization of the heating modes

Choice of optimal high-speed modes and their parameters, of an optimal mode of reductions

Stabilization of rolling modes

Algorithm choice for differential metal heating in furnaces

Choice of algorithms for control of the rolling temperature and speed modes

Choice of algorithms for optimal metal heating in furnaces

Choice of algorithms to control the reductions

Synthesis of algorithms for SUET

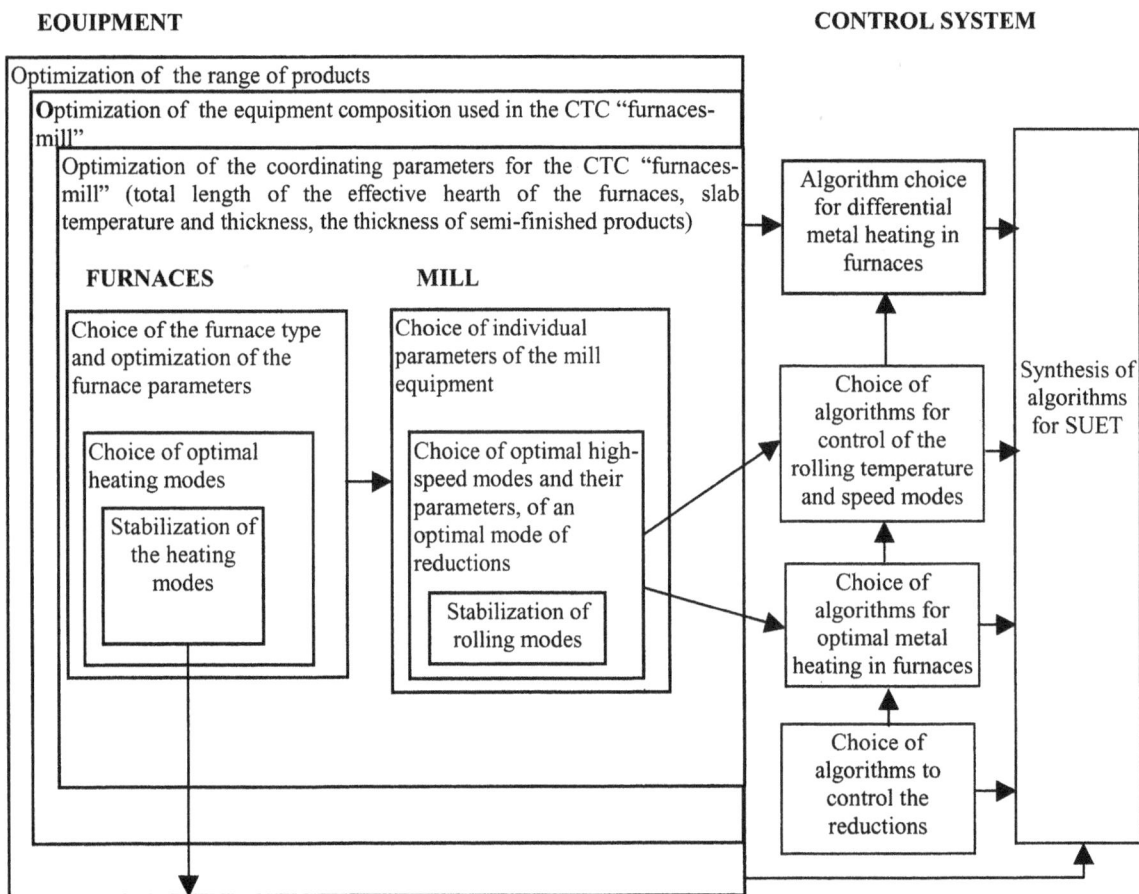

Fig. 1. Reconstruction activities and control algorithms needed to optimize the technical and economic parameters of strip rolling mills

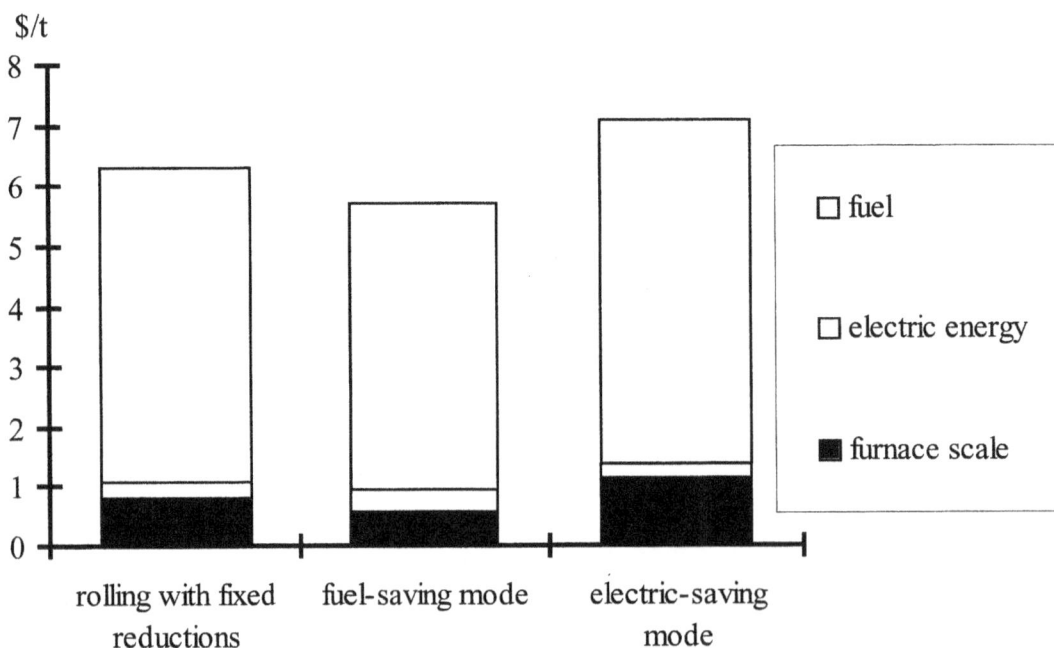

$/t

8
7
6
5
4
3
2
1
0

rolling with fixed reductions   fuel-saving mode   electric-saving mode

□ fuel

□ electric energy

■ furnace scale

Fig. 2. Resource & energy consumption

66

IFAC

Publications
www.elsevier.com/locate/ifac

A SOCIALLY APPROPRIATE APPROACH
FOR MANAGING TECHNOLOGICAL CHANGE

Mico Jancev * and Janko Cernetic *

* Management Consultant (SMILE IT Solutions), Krausegasse 4 – 6/9/5,
AT - 1110 Wien, Austria, E-mail: mjancev@hotmail.com
** J. Stefan Institute, Jamova 39, SI - 1102 Ljubljana, Slovenia,
E-mail: janko.cernetic@ijs.si

Abstract. This paper presents an example of a less known socially appropriate approach
for managing technological change that is able to empower individuals, teams and even
larger social groups to become more competent in coping with changes and consequent
social instabilities brought about by the introduction of new technology. An original
approach and methodology, both called COPIS, are shortly described that can serve for
teaching people the required Change-Management knowledge in the form of typical
workshops and through performing practical projects. The reported approach has been
successfully used in the post-socialist countries of the Balkan region. *Copyright* ©
2001 IFAC

Keywords: International stability, Complex systems, Human factors, Socio-technical
systems, Social impact of automation, Developing countries.

1. INTRODUCTION

Changes in living systems, either on a micro scale
(living cell, organ or living creature) or a macro scale
(organisation, town, country or society), are
unavoidable and cannot be postponed. Each change
brings with it a minor or greater instability to the
systems that are affected by it. Every time in the
development of civilisation when a new technology
appeared, it has come to revolutionary changes. Just
in these times, such technological revolutions are
greatly affecting our lives through electronics,
computerisation and other high technologies. All
these changes bring with them more or less profound
problems to the individuals, organisations and even
to societies, leading to different kinds of social
instabilities that are merely added to those caused by
other reasons, such as political or economical.

Humankind has always sought for means of either
avoiding or coping with changes and the consequent
instabilities. In the case of current technological

changes, avoidance does not appear to be a wise
strategy. Instead, any individual, organisation or
country willing to survive now and in the future, has
to find his own ways for coping with this headline of
modern times.

In recent decades, many more or less successful
approaches have been invented and tried to make the
transition to new technology easier for its developers
as well as for its users. Very broadly, such
approaches may be classified into two wider groups.
On one side, there are the **technically oriented**
approaches for better technological transition
management, such as Systems Engineering, Software
Engineering, Project Management. On the other side,
there is a less popular group of the so-called "**soft-
oriented**" approaches, such as Soft Systems
Methodology (Checkland and Scholes, 1990),
Human-Centred Systems Design (Gill, 1996), User-
Centred Design of Systems (Noyes and Baber, 1999)
and still some other less known approaches. It can be
said that the technically oriented approaches have

laid very good foundations to stabilize the very demanding processes of developing advanced (mostly computer-based) technology. Partially building on that foundations, the soft-oriented approaches have added the consideration of the subtle human and other soft aspects mostly not dealt with explicitly in the technically oriented approaches.

Unfortunately, most of the mentioned approaches leave to their users the efforts of adapting the methodology to the, let say, "social parameters" of the environment where the approach is to be implemented: e.g. language, culture, social norms, and similar. While this challenge may be considered to remain open for some time in the future, the current trends of globalisation indicate that these social issues of technology management and their impact on social stability, particularly in developing countries are promising candidates for further research. In other words: there is a need for more "socially appropriate" approaches for management of technology. The aim of this paper is to make a useful contribution to that direction of research.

2. NEED FOR SOCIALLY APPROPRIATE APPROACHES

Here some reasons from the social research of technology are discussed, in order to support the statement of need for more socially appropriate change-management approaches, particularly for technology management. This more or less theoretical background is mostly based on ideas presented by Westrum (1991). According to this source, there are at least two serious reasons why modern change-management approaches should be socially appropriate. One reason is the essential nature of technology and the other reason is tight connection between technology, economy and society (Fig. 1).

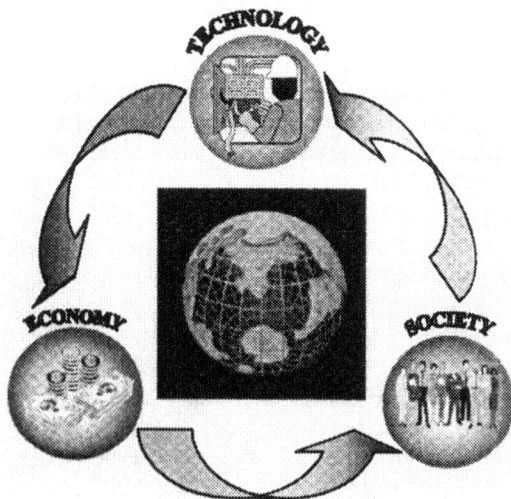

Fig 1. Technology – Economy – Society, three interdependent systems

Westrum (1991, p. 7) defines technology as consisting "of those material objects, techniques and knowledge that allow human beings to transform and control the inanimate world". Further, he explains the three essential elements of technology. For the purpose of explanation here, the first original explanation statement of Westrum is expanded (in italics) by a short reference to the most significant agents of technology.

1. "Technology is things ..." *invented, constructed and used by people.*
2. "Technology is techniques ..." *learned and practiced by people.*
3. "Technology is abstract knowledge ..." *possessed and managed by people.*

The above expanded explanations lead to the second reason why modern change-management approaches should be socially appropriate. This one can be expressed in a number of equivalent ways (Westrum, 1991, pp. 4-8):

- "Technology and society are interacting systems."
- "Technology and society are forces that together shape the world in which we live ..."
- "Technology and society intertwine: when one changes, the other is likely to change as well."
- "Technologies are not isolated things; rather, they are parts of systems of human action."
- "Technology is embedded in a social matrix."

In continuation, the same author comments some instances of how technology is embedded in social life:

1. The social division of labor.
2. Exclusive possession of technologies.
3. Responsibility for operation and development of technology.
4. Association of social groups with technologies.
5. Dependence of social forms on technology.

Taking all the above as acceptable arguments, the reader may ask what is characterictic for a socially appropriate change-management approach to support the introduction of technology. Some ideas are provided below to illustrate this issue that may be more deeply elaborated at another occasion. The following ideal features may be considered when evaluating an approach for its social compatibility:

- does the approach take into account the basic human and social values when dealing with the needs of prospective users of technology?
- does the approach consider the culture of the organisation and of the country when

designing, selecting or introducing the new technology?

- does the approach assure that the new technology will support the user organisation in achieving not only its economic results but also its mission and meaning in relation to its wider social and global context? (e.g. according to the ideas contained in the OSTO model of open socio-technical and economic system proposed by Hanna, 1988; cited from Brandt et al., 2001)
- does the approach provide procedures for advanced analysis as well post-implementation monitoring (measuring) of any possible impacts of technology on its users, the user organisation, immediate society and natural environment?
- does the approach emphasise high standards in professionalism and ethics?
- does the approach encourage or support inter- and trans-disciplinary cooperation?

The above list of "socially desirable" features is not very exhaustive but it can be summarised by a single general sentence: the socially appropriate approach for management of technology should find a dynamic balance in dealing with the business (economic), technical, human and social aspects of technology to be designed, developed or introduced. The term "dynamic balance" means that the approach is continuously adapting to the changes in relevant parameters, such as those in technology, objectives of the user organisation, human (users') needs, social situation, etc.

It may be noted here that some business organisations are able to progress successfully towards meeting many of the above-mentioned "criteria" for social compatibility, just by aiming at a higher level of (business) excellence. In this context, it is interesting to compare how the pretty well known European model of excellence fits to these criteria. The original text says very succinctly that: "The (EFQM) Model, which recognises there are many approaches to achieving sustainable excellence in all aspects of performance, is based on the premise that: **Excellent results with respect to Performance, Customers, People and Society are achieved through Partnerships and Resources, and Processes."** (EFQM, 2001).

3. AN EXAMPLE OF A SOCIALLY APPROPRIATE APPROACH

The COPIS approach and methodology are brifley presented here as a less known example, in order to illustrate how a socially appropriate approach may look like. The reader may understand that this example has a specific social character, as it has been developed in a specific working and social environment. While it is clear that the COPIS approach and methodology can be most successfully used in social ambients similar to those where they were developed, they can serve as a source of useful ideas and principles to be adopted in other approaches that will be appropriate in different social circumstances.

The acronym COPIS explains in the original language of its author that the approach includes a step-by-step Holistic, Meaningful, Professional, Information technology-supported and quantitative analysis of the organisation, based on its problems, its goals and the needs for changing.

Basically, COPIS can be classified into the wider group of change-management approaches. In comparison with some other soft-oriented approaches, the COPIS approach can be characterised as being socially appropriate as well as human-centred. Further, the COPIS approach and methodology have many generic elements that qualify it for use in areas beyond technology management, e.g. in problem areas where there is a need to improve any other aspects of (social) stability.

The COPIS approach and methodology have been successfully used in areas of business improvement, process re-engineering, quality management and technology management (Jancev, 2000-a). Based on partly original psychological, sociological and management concepts, this approach can be applied for managing the introduction as well as the implementation of advanced technology, particularly in situations of greater complexity and during the introduction of new technology in developing countries (Cernetic and Jancev, 2001).

The approach is based on a **very pragmatic** way of thinking and acting. It was developed from practical managerial experience of the first author in a high-tech military-owned organisation involved in manufacturing and maintenance of jet engines. Although staying pragmatical, the approach has integrated many practice-proven elements from the organisational, social and economic science. In the following, some of its basic premises are explained (Fig 2).

First, the approach recognises the **managers** of the organisation as the first line of competent persons who are able to solve in a most efficient way top-level problems of their business. At the same time, the managers are the first to be responsible for problems related to the introduction of advanced technology.

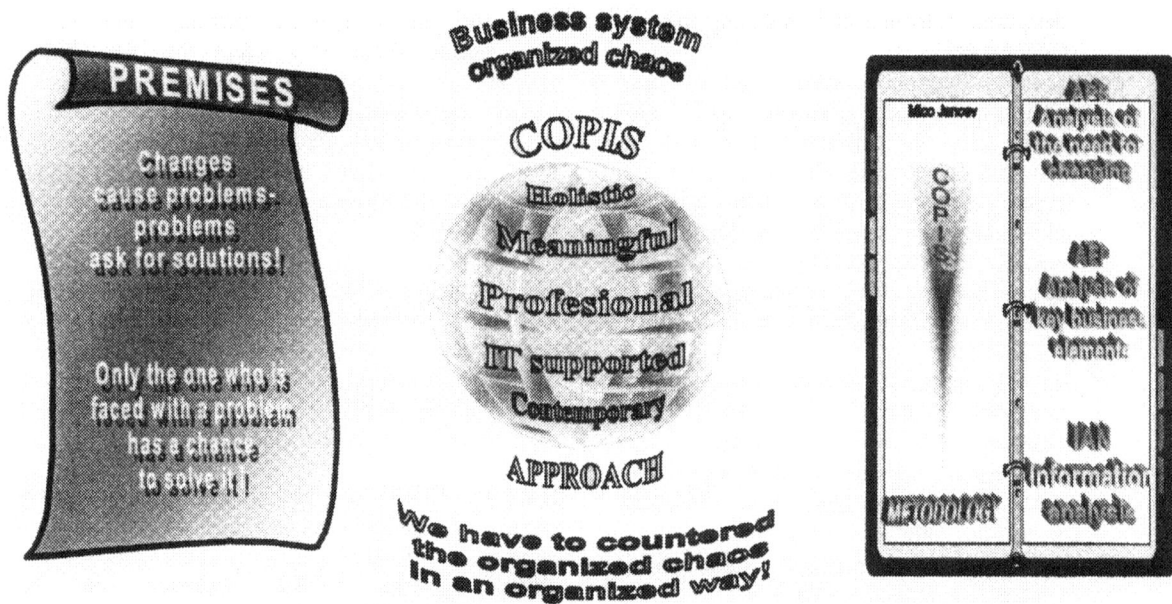

Fig. 2. Short introduction of the COPIS approach.

Second, the problems that occur in a contemporary organisation shall never be observed from only one isolated discipline. For example, almost any problem identified in a modern production-oriented firm can rarely be solved only from the engineering viewpoint; it mostly has economic, financial, organisational, information-technology and people-related implications. Therefore, without a complete and **interdisciplinary problem analysis**, the chances are small to find an appropriate solution that will be approved in practice.

Third, the business system of a production-based organisation, by its nature, is very complex. Its **complexity** is measured, firstly, by a considerable number of its business elements, and secondly, by numerous relations between these business elements. For example, an analysis of organisations where the COPIS approach was implemented has revealed that the number of important business elements in different systems (including electronics, mining, machine manufacturing, aviation, navy, medicinal equipment production and tobacco industry) ranges from about 200.000, up to over 4 millions (Jancev, 2000-b). Controlling such complex systems involves the need to manage an enormous amount of data. The previously mentioned analysis has found that between five and 70 million single data elements are required, depending on industry type.

Faced with such a huge number of business elements, the industrial managers experience daily a situation that professional analysts in America have called "**organised chaos**". Therefore, managing such systems is a very serious task. Thanks to modern technology, this task can be made more manageable. Nevertheless, introducing new technology in such systems must be planned much more carefully than

two decades before. For this purpose the COPIS methodology has been developed.

The COPIS methodology, based on the premises of the COPIS approach, consists of clear phases, steps and procedures (See Fig. 3 at the end of the paper). These help managers and employees in an organisation to work in a systematic, pragmatic and inspiring way, in order to:

- **identify key problems** faced by the organisation, in particular those potentially to be solved by advanced technology;
- **understand problems** by enabling the affected people to discuss them within the interdisciplinary team;
- **define the most important problems**, in order to be able then to
- **begin with problem-solving** through an organised sequence of well-planned projects; and last, but not least, to
- **prepare the social setting** in the organisation in advance for accepting the changes.

The COPIS methodology is guiding the work on projects through **three main phases**:
1. Analysis of the needs for change;
2. Analysis of business elements;
3. Information analysis of the organisation.

Each of these main phases is performed within the project team by means of about **30 pre-defined questionnaires** and over 30 overview-tables. Each questionnaire includes over 300 specific questions and is worked through in two passes. In the first pass, each team member fills in the questionnaires alone. In the second pass, the results of questionnaires are co-ordinated jointly with the team.

In addition to the usual forms of communication, the COPIS methodology includes a programmed series of **teamwork events**. Each of these events offers to the participants of the change-management process an opportunity to meet in an exciting and socially appropriate work setting, in order to discuss problems and to work together creatively. This series of teamwork events includes anything from short one-hour meetings up to 3-day workshops, seminars and dedicated project-management meetings. By these meetings, the participants are guided successively from initially making better acquaintance of each other, through team building and problem analysis, towards negotiating and elaborating desired solutions. Probably the most important and the most typical event in this methodology is the 2,5-day **introductory workshop** where the team starts to "search for the missing link" in the context of the current situation.

4. SOME RESULTS AND EXPERIENCE

The problems and projects dealt with by the holistic COPIS methodology included many relevant business aspects of different organisations. It has helped many organisations in the area of previous Yugoslavia to implement automation and information systems as well as other types of advanced technology in a profitable and »human-centred« way.

Up to now, the COPIS approach and methodology were used in managing **over 50 projects** for about 20 organisations in the area of former Yugoslavia. In over 200 various workshops and seminars about one thousand people took part, having 15 different professions and coming from about one hundred organisations or institutions.

The results of using the COPIS approach are very favourable in many aspects: measurable and non-measurable. Firstly, the organisations that have accepted this approach and methodology succeeded to *shorten the time* needed to introduce new technology, irrespective of their basic activity. For example, the introduction of information technology was, on the average, twice as fast as in other organisations from the same geographic area.

Secondly, one of the most important effects of implementing the COPIS approach was the *deeper insight* into the essential business problems of the organisation. At the same time, the managers, being actively involved into the problem-identification and problem-solving process became more quickly *aware of the possibilities offered* by advanced technology. This means that, in many cases, the benefits of automation or IT are essentially clear before any

(eventually lengthy) feasibility or economic-justification studies are finished.

Thirdly, the mentioned introductory workshop represents an enormously favourable occasion for *building an efficient team*. Such a team becomes a problem-solving community that is really "glued together" by means of making personal relations deeper than it is usually experienced on such occasions, at least in the countries concerned. According to the COPIS methodology, such deeper social bonding of the team is considered to be one of the important ingredients for success in building the previously mentioned "environment for change". Here the soft human and social factors play an immensely important role on levels beyond direct rational consideration (Jancev, 2000).

5. CONCLUSIONS

In this paper the discussion is based on the assumption that changes, in many cases those triggered by the introduction of new technology, are often a reason for social instability. Therefore it is suggested to use the existing change-management knowledge in the area of moderating the tensions occurring during the technological switchover. But it important to do this in a "socially appropriate" way if one wants to assure that "technology will not simply occur but will be carefully chosen and designed to bring about as far as possible only the intended results" (a paraphrased sentence from Westrum, 1991).

Learning themselves from theory as well as from practical work in the post-socialist countries of the Balkan region, the authors of this paper believe that the most important achievements in a particular society are possible through the changes in philosophy, in the way of thinking and in the way of acting. Above all, real changes are possible by shifts in the world-view of each individual and that of the society as a whole. In managing technological change, like in many other areas of human activity, this is possible to achieve by approaches that are socially appropriate in the sense of the example approach discussed in this paper. The best result of working by such an approach is the situation when the involved person becomes proud of the way he/she performs his/her work.

Therefore the **changing process requires a special social environment that** cannot be imposed or requested, but **has to be carefully created**. Creating such an environment means creating social consciousness about the importance of changes and empowering people for proactive change management. The latter is probably the most essential aim as well as contribution of the COPIS approach presented in this paper.

REFERENCES

Brandt D., Tschiersch I. and Henning K. (2001). The design of human-centred manufacturing systems. Chapter 5 in: *The Design of Manufacturing Systems*, Vol V (Leondes C., Ed.), CRC Press, Boca Raton, p. 5-3.

Cernetic Janko and Mico Jancev (2001), Implementation of advanced technology in post-socialist countries, *Proceedings of the 7th IFAC Symp. on Automated Systems Based on Human Skill*, Aachen, Germany, June 15-17, 2000. Published in 2001 for IFAC by Pergamon.

Checkland Peter and Jim Sholes (1990), *Soft Systems Methodology in Action*, John Wiley & Sons, Chichester, New York.

EFQM (2001). European Federation of Quality Management, web site: http://www.efqm.org/

Gill S. Karamjit, Editor (1996), *Human Machine Symbiosis – The Foundations of Human-Centred Systems Design*, Springer-Verlag, London.

Hanna D. P. (1988). *Designing Organisations for High Performance*, Addison-Wesley Publishers, Reading, MA, USA.

Jancev, M. (2000-a). COPIS approach and methodology - key for solving business problems of an organisation (in Slovenian). Proceedings of the 9th Annual Conference of Slovenian Federation for Quality, pp. 59-61. Portoroz, Slovenia, November 9-10, 2000.

Jancev, M. (2000-b). *Change Management – a Challenge for Managers* (in Serbo-Croatian, Macedonian and Slovenian, respectively), materials for the introductory workshop, Ljubljana, Slovenia.

Noyes Jan and Chris Baber (1999). *User-Centred Design of Systems*, Springer-Verlag, London.

Westrum R. (1991). Technologies and Society – The Shaping of People and Things. Wadsworth Publishing Company, Belmont, California, USA.

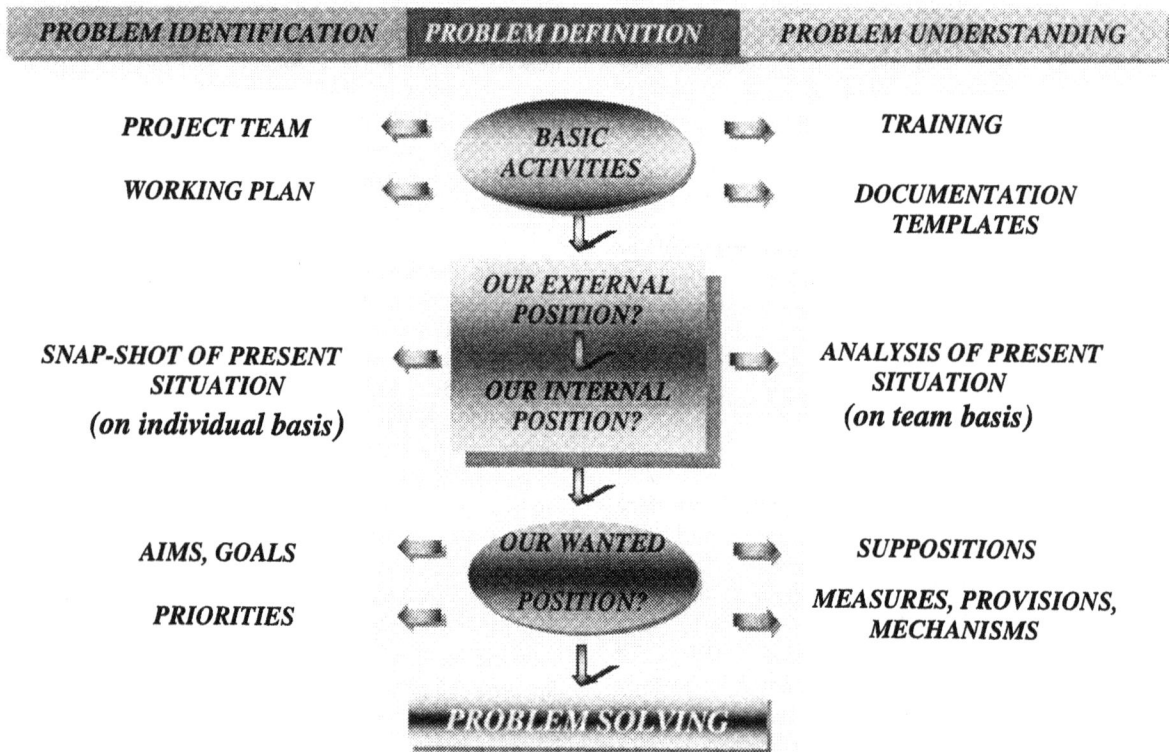

Fig 3. Structure of the initial phase of the COPIS methodology.

IFAC

Publications
www.elsevier.com/locate/ifac

ECONOMIC RECOVERY THROUGH ELECTRONIC MODE 2 KNOWLEDGE PRODUCTION

Larry Stapleton[1], Janko Cernetic[2], Donald MacLean[3], Robert Macintosh[3]

[1]Waterford Institute of Technology, Republic of Ireland
[2]Josef Stefan Institute, Slovenia
[3]University of Glasgow, Scotland

Abstract: This paper presents a vision of engineering and management research as a regional development tool. It also sees these research efforts in a wider context, leveraging economic development and encouraging inter-cultural exchange that is neither imperialistic nor deterministic. It briefly outlines the key elements of the approach and presents some arguments for adopting this approach in western European-Balkan region knowledge production partnerships. *Copyright © 2001 IFAC*

Keywords: Economic Recovery, Knowledge Production, Intellectual Capital, Information Technology

1. INTRODUCTION

The social, political and economic problems associated with the Balkans conflict have been highlighted only too graphically over the past decade. However the region has an internationally recognised academic community that holds important conferences, seminars and publishes widely in the engineering and management disciplines. In modern knowledge-based economy a major key to wealth creation is intellectual capital. It can be argued that leveraging off of intellectual capital can enable a speedy transition to a knowledge-based economy. In response to this broad policy trajectory, countries throughout Europe are increasing their investment in their respective education systems, as a route to medium and long-term growth. The Republic of Ireland strategy of creating an 'e-island' in Europe has been predicated upon a high level of investment in second and third-level education over the past thirty years. Similarly, outside Europe, countries like Singapore have achieved even more success in terms of economic expansion through an advanced, highly skilled work force. Whilst there are many other factors which have enabled small, high growth economies to be successful (including attractive corporate tax laws, social partnership models and so-on), the undoubted role of providing and leveraging intellectual capital has been recognised by policy-makers and has been well documented. This paper argues that a similar, but trans-regional approach which levers off of existing high levels of intellectual

capital, may provide the necessary impetus for economic stabilisation and growth in disadvantaged regions such as the Balkans. It is self-evident that the Balkans has produced some of the leading engineers to be found in Academia, and that there are high levels of intellectual capital in the regions. This paper presents a model of economic development which levers off of existing intellectual capital, and is simultaneously socially responsible and based upon inter-cultural exchange. Few models of economic development exist which combine these important properties with theoretically well-founded knowledge production and knowledge application theory.

2. MODE 2 KNOWLEDGE PRODUCTION

The relationship between academia and practice is currently characterised by a division of labour where academics perform the research to generate knowledge whilst its commercial application is typically led by consultants. Academic research in management is increasingly viewed as irrelevant by practitioners because of the lengthy timescales involved and a tendency to search for generalisable approaches, whilst commercial consultancy is often viewed as promoting the latest management "fad" in ways which are not sympathetic to the specific needs and circumstances of any particular organisation.

In stark contrast with this approach there is a significant new level of opportunity for established form of academic research which focuses on problem-centred research and collaboration with practitioners to engage in real-time generation and application of knowledge on-site. Such approaches are currently being grouped under the term "mode 2 knowledge production."

In the preface to The New Production of Knowledge (Gibbons *et al* (1994)), Michael Gibbons draws the readers attention toward a new form of knowledge production (mode 2) which, although originally an outgrowth from its traditional counterpart (mode 1), is becoming increasingly distinctive.

"Our view is that while Mode 2 may not be replacing Mode 1, Mode 2 is different from Mode 1 – in nearly every respect ... it is not being institutionalised primarily within university structures ... (it) involves the close interaction of many actors throughout the process of knowledge production ...(it) makes use of a wider range of criteria in judging quality control. Overall, the process of knowledge production is becoming more reflexive and affects at the deepest levels what shall count as "good science." (Gibbons *et al*, vii)

Many of these themes are further refined as the key features of mode 2, which are contrasted with the features of mode 1.

"Mode 1 problems are set and solved in a context governed by the, largely academic, interests of a specific community. By contrast Mode 2 is carried out in the context of application. Mode 1 is disciplinary while Mode 2 is transdisciplinary. Mode 1 is characterised by homogeneity, Mode 2 by heterogeneity. Organisationally, Mode 1 is heirarchical and tends to preserve its form, while Mode 2 is more heterarchical and transient. In comparison with Mode 1, Mode 2 is socially accountable and reflexive." (p3)

3. MODE 2 IN OUTLIER REGIONS: EXPERIENCES IN SCOTLAND

Our experiences of using Mode 2 research processes in Scotland indicate that the beneficial outputs can be secured both in terms of business performance and academic insights. Furthermore, the business benefits accrued appear to be of a more lasting and sustainable nature than might be the case with more Mode 1 approaches.

To illustrate this point, we will consider one such Mode 2 project which involved a food manufacturing company located in the west of Scotland. Founded in the early part of this century, the company was run by the third generation of owner-managers and at the time the project began (1996) the company employed

approximately 250 people. The company operated in mature markets, with demand for many of its traditional products experiencing declining demand. A combination of changing consumer preferences, changes in the retail environment and the BSE crisis meant that the company needed rapid and radical change.

The mode 2 project focused initially on an attempt to combine elements of business process re-engineering (BPR) with aspects of organisational learning (OL). Our interaction with the transformation process took the form of intermittent site visits along with hosting off-site meetings, workshops, etc. over a period of approximately 18 months.

Over this 18 month period, the company significantly improved its performance, moving from sizeable losses to reasonable profits in a very competitive marketplace. Turnover increased by 50%, a range of new products were introduced and self-managed project initiatives continued long after the Mode 2 project had concluded. [see MacIntosh and MacLean, 1999 for further details of this case study].

4. INFORMATION TECHNOLOGY & MODE 2 KNOWLEDGE PRODUCTION

Information technology has the proven capability to transform organisational structure and enable highly flexible modes of organisation (Davidow & Malone (1993), March (1999)). Electronic technologies are becoming available which can enable large scale group interaction at a low cost, where members of that group reside in remote regions, even where little or no ground based telecommunications infrastructure exists. A number of information and communications technologies will contribute to any mode 2 knowledge production capability. For example, the following specific technology areas can contribute to inter-regional mode 2 knowledge production:

- Broadband Telecommunications including Satellite Communications
- Internet and Extranets
- Personal and Group communications devices such as cellular phones, PDAs, video conferencing etc.
- Group Asynchronous Communications software including email, notes, and agent-based computing
- Group Synchronous Communications software including chat rooms, telepresence technologies and virtual environments

To date the potential of information and communications technologies for mode 2 knowledge production remains unexplored.

5. STATE OF THE ART

Each of the above technology areas have undergone massive development in the past fifteen years. However, it is has been evident for some time that we continue to have a poor understanding of their impact upon social interaction and, by implication, organisational learning and knowledge production (Sproull and Kiesler (1991), March (1999)). There remain very deep concerns about what these technologies, on their present trajectory, will deliver to society. For example, the French philosopher Jean Baudrillard speaks of the burgeoning obesity of modern computer-based information systems which seek to gather into themselves all possible codifiable knowledge but which simultaneously leave their users less and less informed (Baudrillard (1990)). This 'fatal strategy' in IS and IT cannot deliver useful knowledge indefinitely. The context in which knowledge and information are enacted is lost in the increasing obesity of modern IT. It is also lost as representation and storage technologies increasingly fragment and decontextualise codified knowledge into databases (Davidson (2000)). It is vital that the context of technology is recognised as part of the technology itself. As Ihde (1998) points out,

'cultural embeddedness is a matter of a technology-in-a-context where it *is* what it is only contextually. This is why there is no such thing as a simple technology transfer. There is only culture-technology transfer'. p. 48

Contemporary projects of technology transfer within and between European regions decontextualises technology – technology is removed from the context in which knowledge can be produced. This approach is flawed and both technology and its context must be re-connected.

Concerns about failures in the IT sector, including large scale business systems and group support systems, continue unabated (Stapleton (2000), Grudin 1994)). These concerns are also reflected in the work on social psychology and organisational learning (Postmes, Spears & Lee (2000), Argyris (1999)). It is recognised elsewhere that many modern information systems deployment projects are perceived by managers to be a significant risk to their business (Davenport (1998)) and may have a serious negative impact upon economic growth in the regions in which they are deployed (Byrne, Ryan and Stapleton (2000)). IS research and practise must deliver specific solutions which are grounded in specific application and problem contexts, rather than artificially imposed into organisational contexts with often traumatic results (Stapleton (2000), Cernetic & Jancev (2000)).

The emphasis in Mode 2 approaches on contextually specific knowledge production offers unparalleled promise as a means of overcoming the problems outlined above. Under "e-mode 2", the issue is no longer knowledge or technology transfer; instead the issue is real-time production and consumption of knowledge in a given e-context, unhindered by the notion of transfer between different geographical locations.

6. E-MODE2

In e-Mode2, technologies are deployed to mediate knowledge production, rather than as a store and retrieval mechanism for that knowledge. In this view IT is not primarily an information store and processor. Rather IT facilitates knowledge production, the generated knowledge becoming stored in the organisation itself. Here information storage and retrieval systems are only relevant in so far as they provide support for mode2 knowledge production. Applications that do not help groups produce knowledge become irrelevant. The focus in eMode2 is upon the interconnections and interactions of humans, rather than technologies. Emode2 recognises the nature of communications networks as proposed in Group Process theory. Brown (2000) tells us that

'The network of communication in a group is [a] crucial aspect of group structure. It is helpful to view communication channels in topological terms – as linkages – rather than in units of physical distance' (p.121).

Brown shows us that effective group structure requires an effective communications network in the group. Group processes do not necessarily require physical proximity to be effective. They DO require appropriate linkages in order to be effective. By extrapolation, mode2 can be effective as a group process if appropriate connections are put in place and managed, regardless of physical boundaries and limitations. This is not to say that mode2 will proceed in the same way as it would where players are physically proximal. Indeed, Ihde (1998)'s phenomenology of technology suggests that this is unlikely.

Specifically, the proposed research will aim to develop a electronically mediated version of the features of mode 2 identified above. Work will therefore focus on effecting processes and practices which promote simultaneous occurrence of the following activities in virtual space:

- Trans-diciplinary approach
- Socially districbuted research

- Research problem framed in context of application
- Theory-building & application occurring in process of production

This framework is depicted in figure 1.

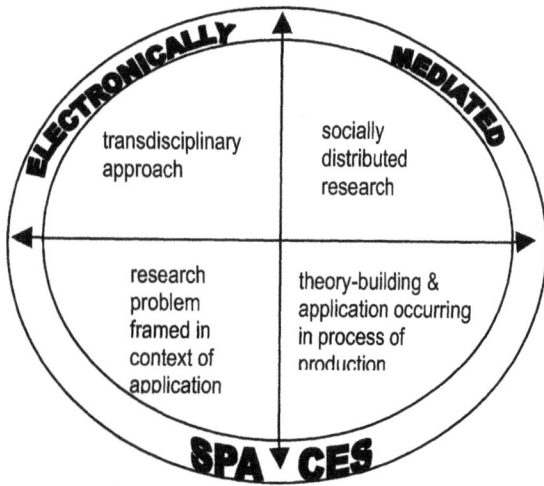

Fig. 1. Electronic Mode 2 Knowledge Production (E-Mode 2)

7. E-MODE2 AND ECONOMIC RECOVERY

This model of social and industrial development encourages Western-European and Balkan joint-partnerships in which cooperative knowledge production delivers industrial growth and economic development. Intercultural exchanges must occur at the level of individual organisational units in e-mode2 and happen in virtual space, eliminating geographical and time-zone barriers. E-Mode2 puts individuals in contact with each other, working towards mutually agreed aims and objectives, negotiated during the initial phase of e-mode 2 knowledge production. This model has been developed utilising concepts adopted in change management methodologies such as COPIS. COPIS has been successfully deployed in over 50 projects throughout the former Yugoslavia and addresses particular conditions in that arena (Cernetic & Jancev (2000), Jancev & Cernetic (2001)). Approaches like COPIS inform critical aspects of the e-mode2 process by addressing, in a post-socialist context:

1. Acquisition of Insight into essential business problems of post-socialist organisations
2. Identification of zones of innovation particular to post-socialist organisations
3. Creation of effective inter-cultural organisational teams, critical for virtual knowledge production

The original mode2 approach as utilised in outlier regions of Europe such as Scotland, already encompasses specific business imperatives found in outlier regions of western Europe. The E-Mode2 approach outlined in this paper focuses upon the specific business and academic issues facing all stakeholders in the mode2 knowledge production process, whether in post-socialist or western European settings. In this way e-mode2 can deliver a socially-accountable approach to knowledge production which is not culturally or economically imperialistic. This mitigates against technological imperialism and exploitation which often accompanies regional conflict or post-colonial settings and which remains too common in the discourse of regional development, particularly in the EU and other western blocs (Banerjee (2001), Faria & Guedes (2001)).

As Jancev & Cernetic (2001) point out, effective socio-economic change in the Balkans is only possible at a very fundamental level and, as Henning (2000) emphasises, effective innovation is most likely at intercultural interstices. E-mode2, as envisioned here, re-centres the change process within the social scene, rather than the technical scene, recognising the unique social settings created by innovative technologies (Stapleton (2001)). The intercultural exchange required must happen in the context of economic recovery, where the practical problems people face are addressed through the inter-cultural exchange mechanisms. This will lead to the new outlooks and relationships required in this troubled region, and help create a 'rising tide which rises all boats'.

8. CONCLUSION

It is evident that a model of knowledge production is required which bring economic recovery to devastated areas and outlier regions, whilst at the same time creating interstices of inter-cultural cooperation 'on the ground'. Management, organisational behaviour and engineering research has not addressed this issue in terms of knowledge production and application, which is fast becoming a major pillar of western economic growth. It can be argued

Through the activities incorporated in the overarching framework that is e-mode2, the devastation of the recent Balkan conflict can be alleviated through inter-community electronic knowledge production which is problem centred, rather than expansionist. The paper presents a theoretical model based upon virtual industrial districts in which organisations are bound together by shared business problems, rather than geography or business sector (as in the case of traditional models of industrial business districts).

REFERENCES

Banerjee, R. (2001). Biodiversity, Biotechnology & Intellectual Property Rights: Unpacking the Violence of 'Sustainable Development', 19th Standing Conference of Organisational Symbolism (SCOS XIX), Dublin, Ireland (forthcoming).

Cernetic, J. & M. Jancev (2000) 'Implementation of Advanced Technology in Post-Socialist Countries', *Proceedings of the 7th Symposium on Automated Systems Based on Human Skill*, Brandt, D. and Cernetic, J. (eds.), International Federation of Automation and Control (IFAC), Elsevier: Amsterdam, pp. 251-254.

Faria, A., A. Guedes (2001). In Search of Less Colonising International Management and Research: Bringing Suppressed and Surprising Experience from Academic Fronts, 19th Standing Conference of Organisational Symbolism (SCOS XIX), Dublin, Ireland (forthcoming).

Gibbons, M, C. Limoges, H. Nowotony, S. Schwartzman, P. Scott, M. Trow, (1994), *The New Production of Knowledge: the dynamics of science and research in contemporary societies*, Sage, London.

Henning, K. (2000). The Aachen Region: Development through Information and Communication, *7th IFAC Symposium on Automation Systems Based on Human Skill: Joint Design of Technology and Organisation*, Brandt, D. & Cernetic, J. (eds.), VDI/VDE-GMA: Dusseldorf, pp. 1-4.

Jancev, M. & J. Cernetic (2001). A Socially Appropriate Approach for Managing Technological Change, *SWIIS 2001*, forthcoming.

MacIntosh, R & D. MacLean (1999), Conditioned Emergence: a dissipative structures approach to transformation. *Strategic Management Journal*, Volume 20, pp 297 - 316.

Stapleton, L. (2001). From Information Systems in Social Settings to Information Systems as Social Settings, *7th IFAC Symposium on Automation Systems Based on Human Skill: Joint Design of Technology and Organisation*, Brandt, D. & Cernetic, J. (eds.), VDI/VDE-GMA: Dusseldorf.

IFAC

Publications
www.elsevier.com/locate/ifac

STABILITY OF SOCIOECONOMIC SYSTEMS: SCENARIO INVESTIGATION METHODOLOGY

Kononov D. A., Kul'ba V. V, Shubin A. N.

Russian Academy of Sciences
Institute of Control Sciences
Tel.: (007)(095)334-90-09
Fax: (007)(095)334-89-59
E-mail: *kulba@ipu.rssi.ru*

Abstract : We introduce a new scientific project, which supposes to create a new effective semiautomatic tool. It's expected to be used for analysis and synthesis of complex social and economical systems. A new principle named «scenario methodology» was developed in order to create this tool. We unite two main concepts: system analysis and the subject-object methodology to describe behavior of the system object. It was constructed formalized event, situation, scenario, principles and methods to define topological structures, create scenario characters, and indicate its properties. There constructed a spectrum of scenario spaces to represent different management fields: technical, technological, organization, economical, lawful and etc. It was defined scenario operations in scenario space. Thus, scenario calculus was created. *Copyright* ©*2001 IFAC*

Keywords : Scenario Methodology, Scenario Stability.

1. INTRODUCTION

Scenarios of development of socioeconomic systems reflect changes in the states and characteristics of their elements and serve as a link between formulation of goals and generation of particular plans of works aimed at preventing and suppressing the after-effects of emergencies. The scenarios are used as the important tools of analysis of the variants of preventive measures for cases of emergency threat, choice of efficient decisions, and coordination of actions to maintain stability. These measures are realized by computerized management systems for prevention and suppression of stability emergencies. The paper proposes models and methods for formalization, generation, and analysis of scenarios of development of socioeconomic systems to investigate the stability phenomenon.

There are two special problems of the decision-making process. The first problem is to represent information as the sequence of events. As a decision-maker chooses the effective control on their professional language, the other problem is to generalize their ideas and knowledge. Representation of sequence of the states as spectrum of scenarios is situated and object-oriented method of data reflection to analyze the problem by natural language. This approach may be used as main tool for effective decision-making and coordination of necessary control system actions.

In control theory the notion named «scenario» is a new notion. Often it is being used in a wrong way because there was not formalized definition. In the paper it will be given mechanism, which if it will be realized permit to synthesize scenario as the tool for semiautomatic analysis of alternative variants of development situation under conditions which decision-maker considers as important limitations.

Formalized notions of the proposed concept of scenario generation were introduced in [0]. The following formal constructs are used: identified system model $M_O(Y;U;P)$, environment model $M_E(X)$, system-behavior model $M_D(Q)$, system-state and environment-state measurement models M_{MO}, M_{ME}, and rules for choosing the profile of object variation $\hat{A}$ (choice model). The set $M=(M_O(Y;U;P); M_E(X); M_D(Q), M_{MO}; M_{ME}; \hat{A})$ of system description will be referred to as the *system metaset*, and its elements will be referred to as the *main elements of the metaset*.

From the point of view of control theory, the special feature of the first model is the presence of the parameters of description of the controlled object, including the vector of phase variables $\mathbf{y} \in Y \subseteq E^m$, vector of controls $\mathbf{u} \in U \subseteq E^r$, and vector of allocated resources $\mathbf{p} \in P \subseteq E^s$.

The main components of the environmental model involve exogenous values to be analyzed in the form of the vector $\mathbf{x} \in X \subseteq E^n$. Within the scope of available information, the decision-makers (DM) make various assumptions about the variations and interdependencies of these values which they cannot control or change arbitrarily. The model $M_E(X)$ is necessary for formal extraction and description of both the exogenous values and their relationships.

The model of system behavior $M_D(Q)$ describes in formal terms the dynamics of variations of the system phase states which is described using the procedures of transformation of the characteristic parameters, as well as the conditions of their interaction with the values describing in formal terms the environment. Here, we formulate the constraints Q that define the conditions for the behavior of the controlled object.

The advisability of extracting the models M_{MO}, $\hat{M}_{ME}$ and $\hat{A}$ was substantiated in [0]. The scenario of variations of the states of an object is a system of models describing the process of variations of its parameters and operation conditions which marks the transitions to new qualitative states that are of importance from the point of view of the researcher. It is advisable to differentiate the object control scenario and object behavior scenario. The former scenario is generated depending on the goal of control and the rule for choosing the control actions, whereas the latter scenario is oriented toward the goals of studying the object. The main difference lies in the fact that the object control scenario includes the subject of control which not only pursues a certain goal, but also actively participates in its realization.

2. MAIN FORMALIZED NOTIONS AND DEFINITIONS

To make a decision, the operating side performs a certain sequence of steps. Let us consider the stages of *monitoring the chosen parameters* and *choosing the control actions* from the standpoint of the DM.

We assume that $\vartheta \in M_0 \subseteq E^d$ is the set of values of the transformation $\Lambda: E^{r+s+l} \to E^d$ which defines the set of all values of the monitored factors.

By the *controllable monitored factors* (CM-factors) are meant the components of the vector $\vartheta = \Lambda(\mathbf{u}, \mathbf{p}, \Delta)$, where $\Delta = [\tau_1, \tau_2]$ is the interval over which the vector functions $\mathbf{u}(t)$ and $\mathbf{p}(t)$ are defined. It is only natural to assume that $U \subseteq M_0 \subseteq E^d$. Therefore, in the course of control the DM follows (monitors, measures) the values ϑ and. if necessary, determines the values $\mathbf{u}(t)$, $\mathbf{p}(t)$, and Δ as the solution of the inverse control problem such as the problem of designing the optimal control. Some factors are generated independently of the operating side and are uncontrollable.

Uncontrollable factors, which include, in particular, the uncertain natural factors, are grouped in the degree of information available to the operating side:

uncertain factors — the vector $\boldsymbol{\alpha} \in N_0 \subseteq E^k$: the operating side knows only the set N_0 of their values, and

random factors — the vector $\boldsymbol{\beta} \in B_0 \subseteq E^l$: the operating side knows the set B_0 of values of the random variable $\boldsymbol{\beta}$: in additional, some information about the law of distribution (that is, the distribution function or probabilistic measure) $\nu(\boldsymbol{\beta})$ of this random variable is known either precisely or to the extent of $\nu(\boldsymbol{\beta}) \in \Omega$, where Ω is the set of distribution laws.

By the *conditional solution* is meant the point $\varsigma = (\vartheta, \boldsymbol{\alpha}, \boldsymbol{\beta})$ of the set $\Gamma_0 = M_0 \times N_0 \times B_0$, and the set Γ_0 itself is called the *set of conditional solutions.*

If there exists a set of alternative variants for achieving the goal, then the DM formulates the priority rule for choosing the CM-factors: the criterion $W_{ef}(\varsigma)$ for operation efficiency. The operating side tries to choose the CM-factors so as to maximize the value of the function $W_{ef}(\varsigma)$.

To construct the scenario of object behavior with the aim of making a control decision, it is advisable that the DM be able to structure the information about uncontrollable factors. The information available to the operating side by the time of making and executing the decision is formulated as an informational hypothesis. This notion is widely used in operations research theory as a means of substantiating the applicability of the *principle of guaranteed result,* which is the basic procedural principle for constructing efficient DM strategies. From the point of view of the goals pursued by the authors of this approach, this was sufficient when the problems of DM information were at the foreground. However, when similar structures are used in control theory, some important factors remain unaccounted for. which makes their detailed description a must. In particular, the problems of structural organization of the object, effect and distribution of the actions, volume of allocated resources, and so on are not considered in precise terms. These factors are the CM-factors or information characterizing the degree of «completion» of the elements of the main metaset. that is, they reflect, in essence, the degree of model completion upon using the concept of incomplete modeling.

By the *quasnnformational hypothesis (QIH) of the DM* will be meant the totality of images of the arbitrary map $\Theta: \Gamma_0 \rightarrow \Gamma_0$; by the time when a decision is made, the operating side knows the construct $\Theta(\varsigma)$, which defines the particular situation in terms of the CM-factors. To make efficient decisions, it is advisable to describe in detail the possible variants of information, methods of organization, assignment of control functions, and so on, that is, to formulate this information in precise terms of informational hypotheses by which the DM abides when analyzing the object behavior and controlling it.

3. FORMALIZATION OF THE ELEMENTS OF A SCENARIO AND STAGES OF ITS CONSTRUCTION

By the *expected event* of object behavior is meant the triple $\Im = (\mathbf{x}(t), \mathbf{y}(t), t)$, where t is the time instant chosen by the rules $A^{(t)}$ and $\mathbf{x}(t)$ and $\mathbf{y}(t)$ are the expected realizations of the parameters of environmental description and phase trajectory, respectively, that were obtained at time t using the models *of M*.

By the *situation* $S(t)$ at time t is meant the chronologized set of events that occurred before t :

$$S(t) = \{\Im^{(i)}(\mathbf{x}^{(i)}(t_i), \mathbf{y}^{(i)}(t_i), t_i) \; 0 \le t_i \le t, \; i = 0, 1, ..., s; \; t_0 = 0\}$$

By the *conditions* $I(t)$ at time t *is* meant the pair $(S(t), \Theta(t))$, where $S(t)$ is the situation at time t and $\Theta(t)$ is the DM QIH.

By the *scenario* $\Re$. *of object behavior* from the DM point of view is meant the sequence of pairs $(I(t_i), t_i)$ that was generated using the choice rules $A^{(t)}$:

$$\Re = \Re\{(I(t_i), t_i)\} \quad i = 0, 1, ..., N; t_0 = 0.$$

N, $T = t_N$, and $\tau_i = t_{i+1} - t_i$, are called, respectively, the *scenario depth, scenario horizon*, and *scenario time step*. Depending on $A^{(t)}$ the time step can be either fixed ($\tau_i = \tau = in\,var(i)$) or variable ($\tau_i = \tau_i(t_i)$), which is defined by the strategies that are used by the operating side to construct the scenario.

The formalized scheme enables one to represent the generation of a scenario at time t_k as a sequence of several stages:

(1) The models M are used to estimate the initial state of the object, that is, refine $\Im^{(k)}(t_k)$.

(2) Some basic state of the metaset is fixed in the form of a possible set of CM-factors, thus, providing the set $M_0(t_k)$.

(3) The situation $S(t_k)$ is estimated, that is, the estimates of the preceding events are updated with regard for the current state of the models *M*.

(4) The current situation is formalized by fixing the current DM QIH $\Theta(t_k)$.

(5) The possibility of continuing the generation of a scenario in the given direction depending on the current state of its main forming elements and environmental conditions of the socioeconomic system (SES) or the need for scenario subdivision (detailing, consideration of fragments, and so on) are established.

(6) The set $M_0(t_k)$ of strategies of scenario generation depending on the formulated QIH is chosen.

(7) The rule of estimation of the scenario generation strategies is refined.

(8) The strategies of generation of the operating-side scenarios depending on the DM information are estimated.

(9) The scenario horizon as the choice of the component $\Delta(t_k)$ of some point $\varsigma = (\vartheta, \alpha, \beta)$ of the set of conditional solutions $\Gamma_0 = M_0 \times N_0 \times B_0$ is updated.

(10) The degree of scenario detailing is refined.

(11) The current event and time step are chosen upon realizing the chosen strategy of scenario generation.

The proposed approach to formalized scenario generation virtually introduces a special *scenario space* $Z^{(SC)}$ whose characteristics and properties were first investigated in [0]. The set-theoretical approach — decomposition of the extended phase space Z

into subsets characterizing the qualitative properties of the object (phenomenon or process) under study that are of importance to the expert — provides a natural way of introducing $Z^{(SC)}$ for the formalized expert's description of operation of the object. Here, experts define the «workspace» $Z^0 \in Z$ where the modeled process is considered.

The expert's description is based on the notions of *expert-important decomposition* (EID) $\varXi$ of the extended phase space Z and *expert-important event* (EIE) occurring sequentially in compliance with the ordering system $\Re(\varXi)$ defined by the objective natural laws (basic model). Analysis of the basic model establishes the basic scenarios of object behavior. In this approach, a diversity of scenarios is ensured by generating different EIDs and systems for EIE ordering, which is the subject matter of the proposed DM QIH. The reader is referred to [0] for the definitions.

The next step in scenario generation consists of establishing the interrelations between the elementary EIDs. The choice of the EIE-effect strategy can be based on various considerations. There exist two basic extreme possibilities: the effect defined by the exogenous expert method *(synergistic approach)* and the endogenous effect based on describing in detail and taking into account the transients — in particular, the transients with a desirable goal — *(attractive approach)*. The entire spectrum of possible variants of models lies between these two extrema. The model of object behavior where the EID $\varXi = \{\{Z^{(\alpha)}\}$, $A^{(e)}\}$ and the effect $\Re(\varXi)$ EIE are defined is called the EID-model. The immediate effect of the EIEs, the path of a given length, the probabilistic characteristics of the EIE effect, and so on, are defined in a natural way. As soon as some QIH is fixed, one can determine the order of the EIEs. Therefore, each point $\mathbf{z} = (\mathbf{x}, \mathbf{y})$ of the extended phase space Z defines, together with the main elements of the metaset, the set of object-behavior scenarios proceeding from the point $\mathbf{z}$.

Scenario may be constructed as system object. Then, we must indicate scenario components, structure and conception. We define set of operations on scenario space and start scenario calculus.

On this way we discover, that the fundamental difference between the synergistic and attractive descriptions of the SES behavior lies in scenario conception (scenario goal).
The attractive description begins with construction of the detailed models $M_O(Y;U;P)$ and $M_D(Q)$, whereas the synergistic description defines them implicitly in terms of the a priori behavior model intro-duced by defining the effect structures (probabilistic, for example). In contrast to the synergistic description, the attractive one is based on analyzing the system behavior by modeling the results of control. This process can have various basic schemes (strategies) that are defined by the DM in the course of a thorough study of the QIH. A convenient classification of the control actions in terms of the informational support, which can be found in the operations research theory, underlies the choice of strategies by the DM. Appropriate methods of the theory of optimal control can be used to advantage to study the controls [0-0].

The attractive description is based on the endogenous determination of its possible variants by detailed structuring of the QIH and studying the strategies of DM behavior, as well as on studying the topological properties of the scenario space. For the attractive description, the schemes of computer-aided generation of the variants of the object-behavior scenario require a detailed description of the information and control components of the QIH and refined synergistic definitions of the EIE effects [0].

4. SCENARIO STABILITY OF THE SES DEVELOPMENT

The applied areas are quite wide: a management to provide ecology security; researching sociological and economical stability system; scheduling and planning; working out effective technology for Production and Logistics; real steps of financial crises; effective strategies for investments and others. Advance imitation system of decision-making supporting based on the scenario methodology has been developed.

We research stability of the behavior scenario of complex systems as application in more time than another one. The most important feature of SESs is their stability (lack of emergencies) in the course of their operation and development. From the most general standpoint, it can be formally characterized as the presence of the following conditions of system operation:
• the possibility of defining and retaining some system parameters within the given domain of the phase space $\mathbf{v} \in Q$, that is, observability and controllability of the system,
• prevention of critical phenomena that wreck the SES elements and/or their relationships (resonance, runaway, and so on), and
• the possibility of ensuring that in the phase space the object moves in the given direction $\overline{\mathbf{e}}$ — for example, in the direction of stable balanced growth — that is, formulation and efficient solution of the problem of deriving, executing, or continuing the system-behavior scenario along the given turn-pike.
•

The notion of stability must be considered not only in connection with the system structure, but also with the possible control actions and behavior of the environment. In terms of the operative use, these controls can be divided verbally into two main groups of static and dynamic controls.

By *static controls* are usually meant the rules that are generated by the DMs and define a finite set, of variations of the SES structure that are weakly related to a particular dynamics of system behavior. In operations research theory, they are called the «passive» control strategies. In essence, the static control actions are conceptual because the expeditious control decisions that are aimed at correcting the SES behavior cannot be studied and realized within their framework. Moreover, in the majority of applications a radical change in the structure of even nonbasic relations is more costly than an operative short-term action upon the system, which can be represented by a pulse fed in some vertex. Owing to these considerations, such controls are used over a fairly large scenario horizon.

Dynamic controls are defined as a infinite, time sequence of structural variations of the SES and external actions on it. The control strategy is realized in time by a sequence of sets of the aforementioned actions supplemented with a time sequence of external pulses fed in the system.

Among the dynamic controls, we highlight the *controls of object-behavior dynamics* that are defined only by the information about the state of the environment and its processes, which was obtained in the course of realizing the previous control decisions and studying the synergistic properties of the object. The totality of the dynamic controls of object behavior can be defined as the strategy of operative control. In this strategy, the control decisions are not predefined but are made on the basis of current information available to the DM, that is, as a rule they are not passive. The instants of decision making are fixed in advance and depend on the current information about the system.

The stability problems were investigated by scenario-connectable points conception.

To analyze the problem we introduce some new definitions: $\Re$-*scenario-connectable points*, $\Re$-*scenario-connectable events*, $\Re$-*scenario-connectable elementary decompositions, connectability and attainability scenario, and expert-absorbing decomposition.*

Let K be an arbitrary set of elementary EIDs. There exists the closed set K that is characterized by the fact that, as soon as the trajectory reaches it, only its events can occur in the scenario. Extraction of the corresponding system structures and the preventive actions or controls retaining closedness relative to the set of EIDs are of great interest for analysis of the SES

and various emergencies. Therefore, when the EID-model is used to study the SES for stability, some set K of EIDs is regarded as the aim of study.

Let us consider some questions which must be answered when studying the EID-model for K-stability:

(1)   Are there closed sets for the EID under consideration?

(2)   Is there a scenario of attainability from the given EID $Z^{(\alpha)}$ for the given closed set K?

(3)   What are the estimates of the attainability time for the given sets of EIDs?

A negative answer to the first question means that the EID-model and object described by it are not stable. In terms of the above definitions, the lack of scenario stability of the EID-model means that it is scenario-irreducible.

Attribute of the Absolute Scenario Instability. The EID-model is absolutely scenario-unstable if and only if there exists an attainability scenario for each pair of the elementary decompositions $(Z^{(\alpha)}, Z^{(\beta)})$ and $(Z^{(\beta)}, Z^{(\alpha)})$ [0,0].

Using the above approach we have got stability conditions for Markov systems. Scenario spaces building on *EID-model on the operator oriented graph* allowed produce the imitations and research the Pulse Stability of SES. The production system describing on von Neumann model was been tested on turnpike stability.

We note that in the description of object behavior in terms of the EID-model the property of scenario connectability means that in one scenario there are the $\mathfrak{I}_{ev}^{(\alpha_1)}(t_1)$ and $\mathfrak{I}_{ev}^{(\alpha_2)}(t_2)$, which are defined, depending on the approach under consideration, by the current EID and the expert probability of the EIE effect, as well as the CM-factor $\vartheta = \Lambda(\mathbf{u}, \mathbf{p}, \Delta)$ used and realization of the conditional decision $\varsigma = (\vartheta, \alpha, \beta)$. In studying the behavior of an object and using one control strategy or another, one needs to establish the repeated events, the frequency of their occurrence, the effect of controls on their occurrence, and so on. These issues are studied within the framework of the notions of «closedness» and «recurrent event» [0, 0].

5.   CONCLUSIONS

The above results have demonstrated the feasibility of formal methods for generation and analysis of the characteristics of SES stability by scenario analysis. The applied areas are quite wide: a management to provide ecology security; researching sociological and economical stability system; scheduling and planning; working out effective technology for Production and Logistics; real steps of financial crises; effective strategies for investments and others.

Advance imitation system of decision-making supporting based on the scenario methodology has been developed. It allowed investigate some models : «Kioto protocol», «500 days» economical conception by G. Yavlinsky, expending of computing technology in Russia, forming the stable regional policy, and others.

REFERENCES

D.A. Kononov, V.V. Kul'ba, S.S. Kovalev-skii, and S.A. Kosyachenko, *Design of Formalized Scenarios and Structural Stability of Complex Systems (Synergism and Attractive Behavior)* [in Russian], Preprint, V.A. Trapeznikov Institute of Control Sciences, Russian Academy of Sciences, Moscow (1998).

D.A. Kononov, S.A. Kosyachenko, V.V. Ku-l'ba, «Analysis of scenarios of development of socioeconomic systems in emergency control systems: models and methods», *Avtom. and Rem. Cont.*, Vol. 60. Part 2. No. 9, 1303-1320, (1999).

D.A. Kononov and V.V. Kul'ba, «Generating development scenarios for macroeconomic processes in terms of the signed graph language,» in: *Modeling of Economic Dynamics: Risk. Optimization, and Forecasting* [in Russian], MGU, Moscow (1997), pp. 7-33.

D.A. Kononov and V.V. Kul'ba, «Ecological management: Object development scenarios and control of the ecological environment,» *Inzh. Ekol.*, No. 6. (1996).

L.I. Rozonoer, «The L.S. Pontryagin maximum principle in the theory of optimal systems. I-III,» *Avtomat. Telemekh.*, No. 10, 1345-1349, No. 11, 1441-1458, No. 12, 1561-1578 (1959).

W. Feller, An Introduction to Probability Theory and Its Applications, Vol. 1, 3rd ed., Wiley, New York (1970); Vol. 2, 2nd ed. (1971).

IFAC

Publications

www.elsevier.com/locate/ifac

HUMANITARIAN DEMINING FOR INTERNATIONAL STABILITY

P. Kopacek

Institute of Handling Devices and Robotics
Vienna University of Technology
Favoritenstrasse 9-11, A- 1040 Wien, Austria
kopacek@ihrt.tuwien.ac.at

Abstract: One of the main tasks of SWIIS is the application of systems- and control engineering methods for international conflict resolution. In the past the classical approaches from control theory, simulation, decision making.... were used. Here a new idea – application of a very well known tool from production automation "advanced robots" - will be presented. A selected field for international stability or conflict resolution is "humanitarian demining", growing up dramatically in the last decade. These robots of the new generation offers possibilities to solve this task in a very efficient way. Finally "Humanitarian Demining Multi Agent Systems – HDMAS" an autonomous, intelligent robot swarm for cleaning minefields is presented. *Copyright ©2001 IFAC*

Keywords: International Stability, Landmines, Demining, Robotics.

1. INTRODUCTION

According to current estimates, more than 100.000.000 anti-personnel and other landmines have been laid in different parts of the world. A similar number exists in stockpiles and it is estimated that about two million new ones are being laid each year. According to recent estimates, mines and other unexploded ordnance are killing between 500 and 800 people, and maiming 2.000 others per month (Red Cross, 1995), mainly innocent civilians who had little or no part in the conflicts for which the mines were laid. Anti-personnel mines are usually designed not to kill, but to inflict horrible injuries instead (McGrath, 1994). However, many victims eventually die of their injuries, and suffer a long and agonizing death, often with little medical attention.

Some countries have banned the use of landmines and others are supportive of a complete ban. However, their low cost ($3 - $30) and the large numbers in existing stockpiles make them an attractive weapon for insurgency groups which operate in may countries with internal conflicts – the most common cause of wars today. They are used for self-defence by villages and groups of people travelling in many districts where civil law and order provide little effective protection. Many countries retain massive landmine barriers on their borders or near military installations. Some of the most severe landmine problems exist in Egypt, Angola, Afghanistan, Rwanda, Bosnia, Cambodia, Laos, Kuwait, Iraq, Chechnya, Kashmir, Somalia, Sudan, Ethiopia, Mozambique and the Falkland Islands.

2. LANDMINES

Landmines are usually very simple devices which are readily manufactured anywhere. There are two basic types of mines:

- anti-vehicle or anti-tank (AT) mines and
- anti-personnel (AP) mines.

AT mines are comparatively large (0.8 – 4 kg explosive), usually laid in unsealed roads or potholes, and detonate which a vehicle drives over one. They are typically activated by force (>100 kg), magnetic influence or remote control.

AP mines are much smaller (80-250g explosive, 7-15cm diameter) and are usually activated by force (3-20kg) or tripwires. There are over 700 known types with may different designs and actuation mechanisms. There are two main categories of AP mines. A blast mine is usually small and detonates which a person steps on it: the shoe and foot is destroyed and fragments of bone blast upwards destroying the leg. When a fragmentation mine explodes, metal fragments are propelled out at high velocity causing death or serious injuries to a radius of 30 or even 100 metres, and penetrating up to several millimetres of steel if close enough. Simple fragmentation mines are installed on knee high wooden posts and activated by tripwires (stake mines). Another common type of fragmentation mine (a bounding mine) is buried in the ground. When activated, it jumps up before exploding. Mines of one type have often been laid in combination with another type to make clearance more difficult: stake mines with tripwires may have buried blast mines placed around them.

3. DEMINING; STATE OF THE ART.

First you have to find the mines and then you must destroy or – today used methods for identifying mines are:
- High-tech methods for mine detection:
  Radar, infrared, magnetic tools, touch sensors (piezo resistive sensor) and so on. Also GPS is used for finding the place again where a mine lies, and for the navigation of the robots.
- Dogs:
  using dogs that sniff the explosive contents of the mines, has significant limitations and cannot be regarded to as general-purpose solution.

Today used methods for destroying and removal are:
- Brutal force methods (include ploughs, rakes, heavy rolls, flails mounted on tanks)
  The problems with this methods are that:
  - ploughs only can by used to clear roads for military use. Mines are only pushed to the side of the road. Some ploughs also attempt to sieve the mines from the displaced soil.
  - Flails are mechanical devices, which repeatedly beat the ground, typically with lengths of chain. These chains are attached to a rotating drum and their impact on the ground causes the mines to explode, but this can cause severe damage to cultivable land.
  - Rollers generally consists of a number of heavy circular disc, which are rolled along

the ground in order to cause the explosion of any mines.

Before demining can start, surveys are needed to produce detailed maps of minefields to be cleared. The survey team may use specially trained dogs to narrow down the limits of a mined area, and normally verifies a one or two metre wide "safe lane" around each minefield to define the minefield which may be surrounded with unknown land or other minefields. Typical minefields are 100-200m across and 0.1-10ha in area.

Hand-prodding is today the most reliable method of mine clearing, but it is a very slow, and extremely dangerous method of mine clearing. A person performing this type of clearing can normally only perform this task for twenty minutes before requiring a rest. This method clears one square meter of land in approximately 4 minutes.

The tools of a deminer are:
1. A whisker wire which is gently swung or lifted gently to check for tripwires.
2. A metal detector which is swung from side to side to check for metal objects.
3. A prodder (typically a bayonet, screw driver or knife) which is used to probe the ground at an angle of about 30 degrees to the horizontal and to excavate earth from around a suspect object. Usually a prodder is used to investigate a suspect metal object. However, which dealing with minimum metal mines or large numbers of metal fragments, the entire area has to be prodded by hand.

One must not forget the essential human skills which deminers need. With experience and training, their eyes reveal vital cues such as slight depressions in the ground caused by settling after mines were buried, their ears can distinguish different sound from the metal detector, and their hands develop a feel for different buried objects.

The UN estimates the cost of removing a single mine at 300 to 1000 $. The primary factor is the local cost of labour. So in low labour-cost countries such as (Cambodia, Afghanistan, or Africa) US$ 100 per month is a high rate pay for manual work, even with the obvious risks. In contrast, the labour cost for de-mining in the former Yugoslavia may be twenty times higher.

Thinking about the number of mines is rather pointless which estimates range from a few million word-wide (including national borders) to 150 million. It is much more sensible to think in terms of the areas of land which are:
  a) known to be affected by mines, and are important to local or displaced population:

homes, food producing land, roads, infrastructure (roads, canals, power lines, water supplies etc.)
b) believed to be affected by mines
c) known or believed to be affected by mines, but land is of no immediate importance.

The standard which is required for humanitarian demining is a guaranteed 99.6% clearance!
So you see there is still a risk of 0.4% to be injured or killed by a mine.

Mechanical mine clearance means either actuating the mine, or removing it for later destruction. Mine actuators imitate the target by hitting the mine, hoping to exert force on the top of the mine at a level above the operating force. The main problem is to find a method of applying the pressure which is relatively immune to the explosive effect of the mine.

When the UN and the supporting armies have to re-establish communications after a conflict, they use the same equipment, but they teleoperate the vehicles to avoid any accident with the foreign operators. This should be named post-conflict demining and not humanitarian demining, because the objectives and the way operations are conducted are completely different.

As breaching vehicles, ploughs are pushed by a tank or an armoured bulldozer. They leave most mines on the side berm they create, where the mines are more difficult to find (this generates additional R&D projects).

There is a bulldozer with a rotating cylinder in front, digging up to 50cm into the ground. The vehicle has been tested in Mozambique. Although it did not reach the 99.6 UN requirements, it removed 25.000 mines in a six-month campaign.

Another demining vehicle uses the same principle, with closer teeth. It is based on a Leopard1 main battle tank chassis to which a rotating roller is added. The tank can be remote controlled from 500m away. In normal terrain this vehicle should clear up to 20.000 square metres per hour with total safety for the mine-clearing team.

4. ROBOTS FOR DEMINING.

Several projects have proposed the use of autonomous robots to search for antipersonnel mines. The sensor problem is nearly solved now and it will take only few time until a combination of sensors will be available and sufficiently tested in order to give full confidence that all the mines have been discovered. There may be false alarms, but no mine must be left. Once the location of a mine is known, several manual techniques are easily applied to neutralize it. A robot can also be developed to do this easy job, which is simply to go to a precise spot, avoiding obstacles and other mine locations. Then it should deposit a shaped charge or some chemical to destroy the mine.

The necessary features of a demining robot are:
- Ability to distinguish mines from false alarms like soil clumps, rocks, bottles and tree roots. This we name high false rate. A high false rate is wasting time.
- Operation in a variety of soil types, moisture contents and compaction states
- Ability to detect both types or in fact variety of different mine types and sizes
- Operation in vegetated ground cover
- Operation on bumpy and/or tilted ground surfaces
- Costs may be lower then 10.000 US$ including the sensors.

Today there are some appropriate, reasonable cheap sensors available or in development based on optical technologies, acoustic and seismic detection, radio frequency resonance absorption spectroscopy, trace explosion detection. Worldwide approximately 100 companies or research institutes offers intelligent, mobile platforms but the price is too high according to the small lot sizes in production. It's only a question of time until this problem is solved.

Random navigation for covering the field and searching for mines has been proposed. Even with improved algorithms applied to a group of robots, it is difficult to accept ignoring a small proportion of uncovered areas. Systematic navigation is theoretically easy with a global positioning system (GPS), but the resolution must be better than the size of the detector, in order to be sure to cover all the area.

A robot has been designed as a light-weight autonomous robot to search for antipersonnel mines. The pressure force on the ground, 5kg, is not intended to trigger the mines. The sensor head oscillates under the alternating movement of the wheels, in order to scan a width of about 1.2 m. the project is suspended until an adequate sensor, weighing less than 4kg, can be installed inside the head.

The Lemming and other small robots studied within the Basic UXO Gathering System (BUGS) programme are interesting because their size may allow them to edge their way inside the dense vegetation of Cambodia and other countries. A systematic coverage of the area to find all mines is difficult to imagine, but with a sensitive TNT sensor, for instance one can imagine the robot exploring the suspected area with a resolution good enough to be able to say that there are no mines around.

Research groups experienced with walking robots try to suggest the use of their devices for this application. Snake robots are more attractive, since they should be able to crawl close to the ground inside dense vegetation. Their design is, however, quite challenging.

The advantages of robots for demining are

- Minefields are dangerous to humans; a robotic solution allows human operators to be physically removed from the hazardous area.
- Robots can be designed not to detonate mines. The advantage is that mines includes a lot of chemicals which when they detonate are put into the ground which is later used for food producing.
- The use of multiple, inexpensive robotized search elements minimises damage due to unexpected exploding mines, and allows the rest of the mission to be carried on by the remaining elements.
- Teams of robots can be connected so that one team is for searching and one for destroying or displacement.

This means that many robots are searching and a few or only one robot is destroying or displacing the mines.

But there are also disadvantages of using robots:
- it is very difficult for robots to operate in different frequently rough terrain
- they are still expensive
- you need special qualified operators

5. HUMANITARIAN DEMINING FOR INTERNATIONAL CONFLICT RESOLUTION

From the systems theoretical and engineering viewpoint SWIIS dealt until now mainly with time continuing systems well known from the field of process automation. Meanwhile in the field of production automation or in terms of systems engineering - time discrete, digital processes - new methods comes up in the last years, probably applicable to the tasks of SWIIS. One of the main ideas of SWIIS in the early eighties was to apply modern methods, developed in systems as well as control engineering for resolution and avoidance of conflicts. More or less conflicts could be seen as a classical stability problem. There are stabilising an de-stabilising parameters. Humanitarian demining is definitely a parameter of the first category. After a conflict – or possibly a war – a minefield occupy land, homes, infrastructure. A lot of organisations worldwide use clearing of minefields and reactivate the land for the displaced local population as an integrating factor in peace discussions to offer native people (e.g. in Kosovo) the possibility to came back to their homes and their lands.

As pointed out demining is today a very time consuming, dangerous and expensive task. Automatised demining e.g. as presented in this paper by robots, is today and will be in the future a powerful tool to solve these problems.

One of these new approaches are "Multi Agent Systems – MAS". These systems are very well known in software engineering since more than 20 years. In the last years there are some works related to the application in production automation. A MAS consists of a number of intelligent, co-operative and communicative hardware agents e.g. robots getting a common task. Because of the intelligence they are able to divide the whole task in subtasks as long as at least one of the agent is able to fulfill one subtask. One practical example is: assemble a car.
Repeating this procedure yields to the solution of the common task. Newest research goes in the direction of MMAS – Multiple Multi Agent Systems – different MAS are involved for the solution of a complex task.

In SWIIS we have some similarities to MAS. Our actors are humans with a distinct degree of intelligence and ability to communicate and cooperate with others. A conflict could be defined as a competition between two or more MAS – in terms of system engineering a MMAS – with not co-operative single MAS. That's definitely a difference to production automation – here we create always co-operative MAS working together and not contraproductive.

This new approach could be an additional step to improve the original idea of SWIIS.

6. SUMMARY AND OUTLOOK.

As pointed out earlier all the existing and planed robots for humanitarian demining are only able to detect the mines. Brutal force methods destroy mines without detection – but with a not reasonable probability. In a next step our robots have to be able to remove the mines from the ground.

The Institute of „Handling Devices and Robotics – IHRT" at Vienna University of Technology is working since 5 years in the field of intelligent, low cost mobile robots. Together with the Austrian industry we developed a „tool kit" for mobile robots adapting such robots for various tasks. This tool kit is based on a mobile platform on which various devices (e.g. robot arms, lifts, sensors,..........) can be attached. With this new concept we have in fact a multipurpose mobile robot for a broad variety of tasks available.

These concepts could be also applicable for " Humanitarian Demining" with minor adaptions. Our tool kit for intelligent, mobile robots offers the possibility to develop, in a very easy and cheap way, deminig robots with a broad variety of features (e.g. different mine detecting sensors, different moving mechanisms, various gripping devices..........).

In a mid or long term perspective it might be possible to develop " Humantarian Demining Multi Agent Systems – HDMAS " consisting of a number of such robots or agents (Kopacek, 2000).

LITERATURE

Kopacek, P. (1999): Preprints of the IARP Workshop on "Humanitarian Demining", Harare, November 1999.

Kopacek, P. (2000): SWIIS – An Important Expression of IFAC's Commitment to Social responsibility, *Preprints of the IFAC Workshop on Supplemental Ways for Improving International Stability – SWIIS* 2000, 22. – 24. May 2000, Ohrid, Macedonia

McGrath, R. (1994): *Landmines, Legacy of Conflict: A Manual for Development Workers*, Oxfam, Oxford.

Red Cross (1995): Landmines must be stopped. International Committee of Red Cross, Geneva, Switzerland.

Trevelyan, J. (1997): Robots and landmines. *Industrial Robot*, **24, Nr.2**, p.114-125.

IFAC

Publications
www.elsevier.com/locate/ifac

LEARNING TO LEARN WITHIN DYNAMIC MULTINATIONAL ENVIRONMENTS: A NEW MODEL

Ted O Keeffe

Centre for Business Excellence, Newtown Hill
Tramore, Co. Waterford, Ireland
E-mail: tokeeffe@wit.ie

Abstract: The realities of global competition and increased customer sophistication have focused organisational attention on the need to develop a "learning culture". However, while much has been written on the importance of evolving a "learning culture", less attention has been given to understanding in a practical way the characteristics of learning organisations and the ways in which companies can improve their learning systems This study of selected multinationals in Ireland identifies organisational learning antecedents and key characteristics that can be located onto a new learning model that is driven by customer and competitive needs, requiring executive management commitment and visible support to be successfully implemented. *Copyright © 2001 IFAC*

Key Words: Learning, Corporate Strategies, Management, Effective, Quality.

1. INTRODUCTION

This paper develops and explores the characteristics of organisational learning, yet these characteristics on their own do not, amalgamate to create learning organisations, they merely stimulate organisations to take better advantage of the status quo. This idea is not universally acceptable, some commentators argue that the aspiration of becoming a learning organisation is not only necessary but essential for long-term survival and the rate at which individuals and organisations learn may become the only sustainable competitive advantage. Seven characteristics of organisational learning have been identified from an extensive review of the literature, which are outlined below.

1. Learning antecedents
2. Environment of innovation
3. Perceived need and learning mechanisms
4. Executive challenge and learning processes
5. Cultural imperative of resourcing learning
6. Organisational wide learning
7. Learning organisation

However, in order to facilitate our understanding of the significance of each characteristic and how it impacts on the learning process the first stage of the learning

model developed is outlined below. The model is designed so that the distance between each coil represents the time required to achieve satisfactory performance in one characteristic before moving onto the next.

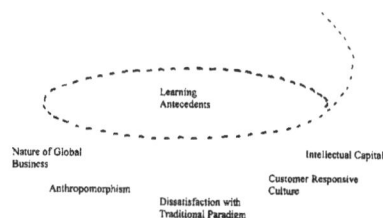

Fig. 1: Learning Antecedents of Organisational Learning

1..1 Customer Responsive Culture

Sinkula (1994) argue the organisation's culture drives the overall value system providing strong norms for sharing of information and reaching consensus. Day (1994) elaborates: "A customer driven learning culture supports the value of thorough market intelligence and the necessity of functionally co-ordinated actions directed at gaining a competitive advantage". With its external emphasis on developing understanding with regard to customers and competitors, the market-driven learning organisation is well positioned to anticipate the developing needs of its customers through the addition of innovative products and services. Thus, a customer focus is an important aspect of a learning orientation.

A narrow construction of culture would lead to learning only within traditional boundaries. To develop a powerful foundation for extensive learning, the organisation must provide the opportunity for generative learning by all stakeholders. Only those constituencies already possessing or developing competitive orientations and company wide learning initiatives have the potential to create superior customer value and competitive advantage. It could be argued that a learning culture is inherently entrepreneurial. A culture that values knowledge and innovation provides the environment in, which learning from exploration and experimentation is most likely to take place (Hamel; Prahalad 1991).

1.2 Anthropomorphism within Organisations

Anthropomorphism is the attribution of human form or qualities to non-human entities. The current, acceptance of organisational learning involves anthropomorphism as the known theory glosses over the, if and how organisations learn, which is far from self-evident. The argument that organisational learning cannot and should not be treated as an extension of individual learning was made by (Argyris & Schon, 1996). Accepting that learning produces knowledge, organisations and their members often know, or come to know, different things.

Although organisational learning occurs through individuals, it would be a mistake to conclude that organisational learning is nothing but the cumulative result of their members' learning. Organisations do not have brains, but they have cognitive systems and memories. As individuals develop their personalities and beliefs over time, so too will organisations develop their views and ideologies. "Organisations select the stimuli to, which they respond because they typically face much more information than they can sensibly process" Hedberg (1983).

1.3 Intellectual Capital

Up until recently manufacturing resources were determined on the basis of capital, land, and business acumen. With the advent of modern technological developments, new management practices and systematic staff development initiatives, the importance of capital and land is diminishing. While at the same time, labour and in particular intellectual labour has generated a very powerful influence across all media to high technology sectors. Graham (1996) stated that for several decades the world's most prescient observers of societal change have predicted the emergence of a new economy in which intellectual prowess, not machine capability would be the critical resource.

If Drucker (1992) has been credited with foretelling the downfall of capitalism, (Peters and Waterman 1982) established the organisational significance of each employee within a company irrespective of size. They suggest that in excellent companies employees be identified as a key resource. Their research showed that "excellent companies were, above all, brilliant on the basics. Tools didn't substitute for thinking. Intellect didn't overpower wisdom. Analysis didn't impede action. Rather, these organisations worked hard to keep things simple in a complex world". They demanded the highest quality and they fawned on their customers. They listened to their employees and treated them like adults. They allowed their innovative product and service "champions" enormous leeway and they allowed some chaos in return for quick action and regular experimentation Peters (1992) argued, "the ability of individuals must be liberated and while strategy, organisation and processes are important, it is the workforce that is the means and driving force of organisational success". Handy (1981) agrees even though people only appear as costs in the balance sheet, they are assets, in that they are or should be a productive resource. People are a resource that needs maintenance and proper utilisation that has a finite life and an output greater than its cost.

1.4 Dissatisfaction with the Traditional Management Paradigm

The study of organisations is multidisciplinary in which learning is a separate element. It is the performance of the overall organisation that determines its survival therefore it is necessary to control it, ensure acceptable standards are maintained and that the necessary corrective action is taken to correct deviations. Drucker (1993) contrasts the notion of need for control in modern organisations with the development of information technology and increasing numbers of 'knowledge employees' who are less likely to respond to autocratic management. Rather he sees employees seeking opportunities for challenge, as well as outlets for creative ability, while enjoying the stimulation of working with like-minded individuals.

This dissatisfaction probably is best articulated by Handy (1989) when he expresses the fear that in "a world where the individual is left even more to his own devices, as the complexities of life and work develop outside the institutions of society, could be a world designed for selfishness". Though Handy's hypothesis is concerned with change in an increasingly complex society and the idea that "the only prediction that will hold true is that no prediction will hold true". Dissatisfaction with the existing paradigm is creating an empty space in the control process, which the concept of organisational learning is endeavouring to fill. The concept of learning and knowledge transfer is attracting greater acceptance because it purports to overcome many of the concerns that have led to the discrediting of the traditional paradigm.

The pace of product, technology and market change continues to accelerate, as learning organisations and competitors become more nimble, flexible and responsive to customers needs. In every industry, these new competitors already have begun to rewrite the rules of effective strategy. As traditional sources of competitive advantage become more fleeting, conventional approaches to formulating and implementing organisational strategy will not provide managers with the tools and insights they need to compete.

1.5 Nature of Global Business

Globalisation doesn't just mean conducting business across national borders. It also means expanding competition for almost every type of organisation. Today's executives must understand that they face foreign competitors as well as local and national ones, Robbins (1997). Such globalisation of multinational organisations presents management with the challenge of learning to operate in diverse cultural settings.

Consumers now have wider choices and are becoming more sophisticated in their selection of products and services. They expect new and improved products, superior service and lower prices. The two major forces driving globalisation have been market growth and cost reduction initiatives. If an organisation wishes to gain market share, expanding operations outside its national borders is one such strategy. In recent years both trade and political barriers have been lowered or eliminated altogether by the creation of multi-country trading blocks.

2. ENVIRONMENT OF INNOVATION

Organisations that operate in fast-changing, competitive environments are pressurised to learn in

order to survive the threats of hostile competitors and difficult environments Toffler (1990). Questions on how and why firms need to be innovative are related to the more general questions of how and why firms differ in practice. However, resource based theorising has typically not been concerned with the practicalities of managing effective, innovative organisations. Their prescriptions draw in varying degrees on theory and empirical observation, though their styles differ markedly. Other writers (O'Keeffe and Harrington, 2001) emphasise the strategic value of managing the acquisition and application of knowledge and thus organisational learning.

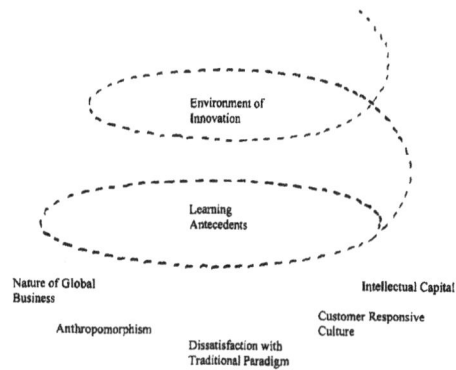

Fig. 2: Learning Model - Phase II

The only sustainable competitive advantage is an organisation's ability to learn faster than its competitors Dixon (1997). Thus knowledge diffusion initiatives become a vital consideration for organisational learning and effectiveness. Recognising and communicating the merits of knowledge is easy; however, difficulties abound whenever organisations begin deliberate organisation wide implementation. While young, dynamic organisations emphasise creativity and innovation as key factors of organisational culture, older, more mature companies may find the implementation of learning initiatives a particularly challenging activity.

Nonaka and Takeuchi (1995) note: "the essence of strategic change lies in developing the organisational capability to acquire, create, accumulate and exploit the knowledge domain". Executive management in particular has a duty to create and communicate knowledge and vision within their organisation. As knowledge creation is synonymous with innovation and new product development, it is an excellent process for creating organisational change. It would appear that innovation and change is best realised as continuing, imaginative and proactive organisational enhancement, not isolated, spasmodic change episodes. A complex environment calls for an integrated workforce, transformational and facilitative leadership.

3 PERCEIVED NEED AND ORGANISATIONAL LEARNING MECHANISMS (OLMS)

All members of an organisation are continuously engaged in learning, helping others to learn and sharing their learning Lawler (1988). Thus, the question of to what degree a particular organisation is a learning organisation can be answered by examining the range of OLMs it regularly utilises, such as benchmarking, innovation and a contingency approach among others. Dodgson (1993:387) defined learning organisations as "firms that purposefully adopt structures and strategies that encourage learning". This idea leads us on to the next phase of the learning model, which examines learning strategies and mechanisms.

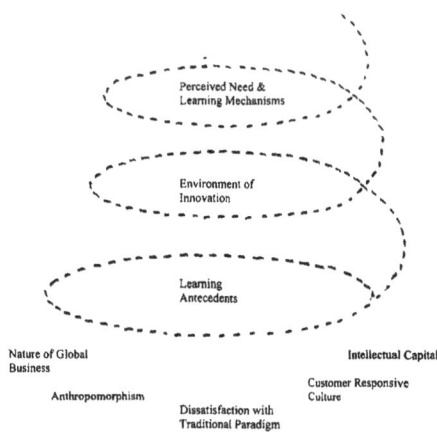

Fig. 3: Learning Model - Phase III

Learning occurs when organisations synthesise and institutionalise people's intellectual capital, memories, culture, knowledge systems, routines and core competencies. Employees may come and go and leadership may change but an organisation's memories preserve behaviour, norms, values and "mental maps" over time. As an organisation addresses and solves problems of survival, it develops a structure that becomes the repository for lessons learned. It also creates core competencies that represent the collective learning of its employees, past and present. As members of the organisation leave and new ones join and are socialised, knowledge and competence are transferred across generations of learning.

4 EXECUTIVE CHALLENGE AND PROCESSES

Learning cannot occur in a vacuum, it requires executive management commitment and functional support in order to develop from discrete to organisational wide activity. This means that all aspects of the company should be actively embracing learning. In learning organisations management decisions are

seen as contingent rather than as definitive but always remain an important part of the strategic decision making process. This allows us to locate the next two phases of the model dealing with the challenges facing executive management and the cultural imperative of resourcing learning.

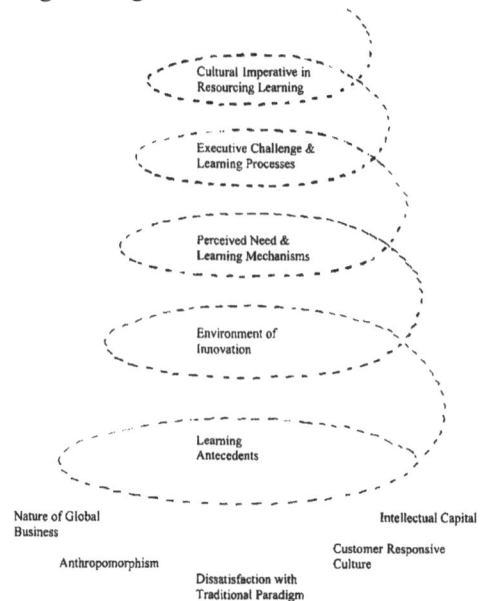

Fig. 4: Learning Model - Phase IV & V

Marsick and Watkins (1999) contend that the juncture between emotional and cognitive aspects of learning poses the biggest challenge to human resource developers seeking to enable effective, continuous learning in the workplace. Argyris (1991) points out that these emotionally charged tasks often are hardest for those that must lead the way. However, organisational development and the imperatives of developing organisational learning are not solely dependent on the competence of executive management, there is a cultural perspective that must be considered as well.

Effective learning is contingent on establishing a culture that promotes inquiry, openness and trust Slocum, et al., (1999). Thus, organisational learning has two facets, a tangible "hardware" facet that consists of learning mechanisms and an intangible "software" facet that consists of shared values and beliefs that ensure that actual learning (i.e., new insights and behaviour) and not mere rituals of learning are developed. According to Schein (1990) organisational culture is a normative system of shared values and beliefs that shape how organisation members feel, think and behave. It is here that corporate values such as respect for individual expression and operating principles are formally espoused within a context that inspires superior collective performance while simultaneously reducing autocratic authority Graham (1996). Hamel and Prahalad (1994) argue, "Any organisation that cannot imagine the future won't be around to enjoy it". Before continuing with the discussion on organisational wide learning it might be appropriate to place this element of the learning characteristics onto the learning model.

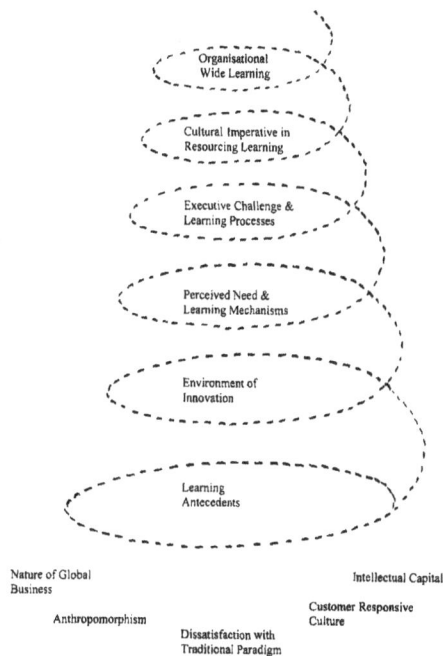

Fig. 5: Learning Model - Phase VI

5 ORGANISATIONAL WIDE LEARNING

The learning process demands unlearning as much as learning. Unlearning involves the process of restructuring past successes to fit the changing environmental and situational conditions. Mistakes, failures, environmental uncertainty or poor performance frequently trigger it. Under any circumstances, unlearning is difficult. Unlearning generates innovation, improvisation and experimentation in new ways of doing things. It leads to the creative processes of learning, change and strategy development. Organisational learning differs from individual learning in several important respects Stata (1989). First, it is a collective event. As a result, organisations learn only as fast as the slowest link. Learning is not simply the sum of each member's learning. Organisational learning must blend the mental models of individual executives with those that are shared among teams and groups. Central to the success of any learning programme is the concept of failure. "All learning takes place in the context of failure, if you are learning to do something and it does not

involve failure, you haven't learned anything" Schwenk (1986).. In real life, of course, failure often involves embarrassment and lack of self-esteem.

"The demands of the 21st century will require business organisations to become more customer focused, using employee talent to create, share and utilise information as part of a broad-based competitive strategy. Another part of this transition will see organisations undergoing significant structural change, developing horizontal networks of task-focused teams leading to flatter organisational structures. The horizontal organisation will be organised around processes rather than tasks, driven by customer needs and inputs and dependent on team performance" Jacob (1995).

6 LEARNING ORGANISATIONS

"A learning company is an organisation that facilitates the learning of all its members and continuously transforms itself" Pedler et al., (1989). According to Garvin (1993) "it is an organisation skilled at creating, acquiring and transferring knowledge and at transforming itself to reflect new knowledge and insights". This brings us to the central conundrum of the learning organisation, if management can be learned, can learning be managed Jones and Hendry (1994). It is accepted that organisations have the capacity to learn collectively and that such learning occurs at different speeds and levels within companies. Organisations do not have brains but they have cognitive systems and memories. Individuals come and go and leadership change, but organisations' memories preserve certain norms and values. However so long as organisational learning refers exclusively to the "sum total of the organisation working in unison" without reference to expanding and building on that, which remains undeveloped, then the point of the learning organisation concept is being missed. Effective learning requires an organisation to resolve the paradox of how to relax its control over the learning process while channelling the benefits from it. For all its elusiveness, organisational learning is more than a metaphor it is attainable, it has a recognisable and distinctive 'feel'. Dale (1994) captured this understanding when she stated that a learning organisation strives to create values, policies and procedures in which 'learning' and 'working' are synonymous throughout the organisation. That learning is inextricably bound up with organisational change and will seek to develop beyond first order learning towards second order change, that is, learning how to evolve the capacity to continuously generate new insights. Achieving a learning organisation, then, requires activity on a wide range of fronts. According to (Mabey and Salaman, 1998) "it demands serious, far-reaching and probably uncomfortable commitments and change from senior management, penetrating to the very basis of the organisation".

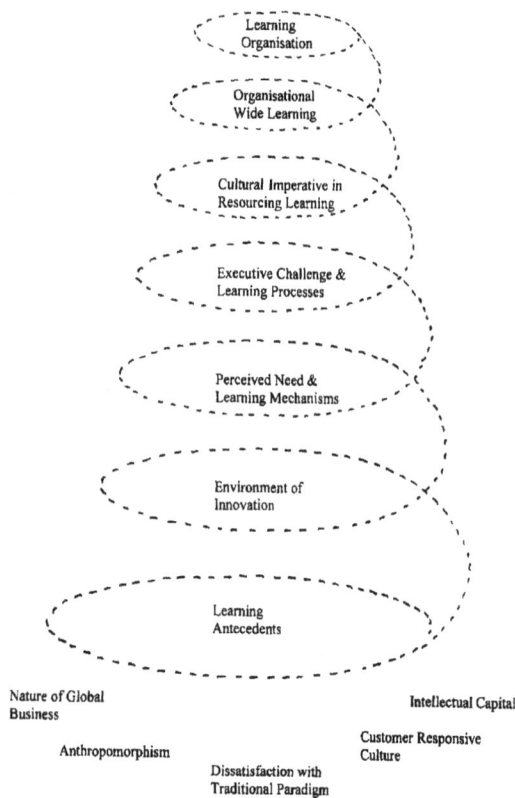

Figure 6: Learning Model - Phase VII

7 SUMMARY

In this paper the building blocks or antecedents of organisational learning were introduced and examined, as were the characteristics of learning organisations that play a prominent part in the development of organisational learning. The move towards a learning organisation begins with the integration of the antecedents together with a convergence of circumstances both internal and external to the organisation. These antecedents are not stand-alone entities; on closer examination one can observe some similarities between them. The increasing importance of knowledge as a source of competitive advantage is closely linked to the move from capital towards knowledge as the key manufacturing resource. The increasing role of intellectual capital in particular and its unique capacity to create knowledge becomes the raison d'être for organisations to develop a capacity for learning. As one's understanding of learning and knowledge creation develops the more one realises the critical function that is undertaken by its people, for it is only through people that an organisation acquires and applies knowledge.

The dissatisfaction with the existing management paradigm is closely linked to an organisation's ability to manage the ever-increasing pace of change and innovation that companies continue to experience Senge (1990). The competitive nature of the business environment with its progressively increasing pressure on companies to respond to customers changing needs is placing undue demands on organisations. The advent of the global market gives customers the opportunity to select products or services from companies from all parts of the globe, which in turn gives customers the opportunity to demand improved performance from their suppliers particularly when alternative quality producers are available.

Where the cultural understanding is focused more on task performance rather than on process and reflection, the ability of the organisation to interpret the vast array of data impacting daily will be severely curtailed. On the other hand where the organisation's approach to developing its resources is comprehensive, inclusive and supportive in a blame free environment its learning capacity will be enhanced accordingly.

These are the foremost characteristics that are driving the trend towards the learning organisation concept, but there are other factors that play a contributing part as well. These include community and moral pressure brought to bear in organisations to develop new processes and procedures to alleviate public concerns while at the same time meeting the new demands of innovations in information technology, which is impacting almost all organisations and changes in society. The growth in information technology is the vanguard for the development of flexible working, the virtual office, outsourcing and temporary employment contracts. The contribution of these factors may not yet be significant but some companies perceive the process of organisational learning as an appropriate response to some or all of these developments.

BIBLIOGRAPHY

Argyris, C., 1991. "Teaching Smart People How to Learn", Harvard Business Review, Vol. 72: May/June, pp. 99-109.

Argyris, C. & D. A. Schon, 1996. Organizational Learning 11, Addison-Wesley Publishing company.

Dale, B. G., 1994. Managing Quality Control: McGraw-Hill, London.

Day, G. S., 1994. "Continuous Learning About Markets", California management Review, Vol. 36: Summer, pp. 9-31.

Dixon, N., 1997. The Hallways of Learning, American Management association, Organisational Dynamics, Vol. 25, No. 4, Spring pp. 23-34.

Dodgson, M., 1993. Technological Collaboration in Industry: Routledge, London.

Drucker, P. F., 1992. Managing For the Future: The 1990s and Beyond. Dutton, New York.

Drucker, P. F., 1993. Post-Capitalist Society: Butterworth-Heinemann, Oxford.

Garvin, D.A., 1993. 'Building a learning organisation'. Harvard Business Review, 71(4): (July/August), pp.78-91.

Graham, A., 1996. The learning organisation: Managing knowledge for business success. The economist intelligence Unit: New York.

Hamel, G. & C.K. Prahalad, 1991 "Corporate imagination and Expeditionary marketing'. Harvard Business School Review, Vol. 69, No. 4: pp. 81-92. July/August, Boston: MA.

Hamel, G. & C. K. Prahalad, 1994. Competing for the Future: Breakthrough Strategies for Seizing Control of Your industry and Creating

the Markets of Tomorrow, Harvard Business School Press, Boston: MA.

Handy, C. B., 1981. Understanding Organisations: Watson & Viney Ltd.

Handy, C. B., 1989. The Age of Unreason: Business Books, London.

Hedberg, B.L.T., 1983. How Organizations Learn and Unlearn: pp. 8-27 in P.C. Nystrom & W.H. Starbuck (Ed.), Handbook of Organizational Design. London: Oxford University Press.

Jacob, R., 1995. "The Struggle to Create an Organzation for the 21st Century", Fortune, April 3, pp. 90-99.

Jones, A. M. & C. Hendry, 1994. "Research Note: The Learning Organization: Adult Learning and Organisational Transformation", British Journal of Management, Vol. 5: pp. 153-162.

Lawler, E.E., 1988. Choosing and Involvement Strategy: Academy of Management Executive, Vol.2: pp. 197-204.

Mabey, C.I.P. & G. Salaman, 1998. Human Resource Management: A Strategic Introduction, Blackwell Business, Oxford UK.

Marsick, V. J. and K. E. Watkins, 1999. Looking again at learning in the learning organisation: a tool that can turn into a weapon, The Learning Organisation Vol. 6, No. 5, pp. 207-211.

Nonaka, I. & H. Takouchi, 1995. The Knowledge-Creating Company: How Japanese Companies Create the Dynamics of Innovation, Oxford University Press, New York.

O'Keeffe, T., & D. Harrington, 2001. Learning to Learn: An Examination of Organisational Learning in Selected Irish Multinationals. Journal of European Industrial Training, MCB University Press, Vol. 25: Number 2/3/4

Pedler, M., T. Boydell & J. Burgoyne, 1989 'Towards the Learning Company.' Management Education and Development, Vol. 20/1: pp. 1-8.

Peters, T. J. & R. H. Waterman, 1982. 'In Search of Excellence Lessons from America's Best-Run Companies, Harper & Row: Publishers, New York.

Peters, T. J. 1992. There's More to a Successful Company than Zero Defects: San Jose Mercury News, July 20: 2F.

Robbins, S. P., 1997. 'Managing today': Prentice-Hall, Inc.

Schein, E.H., 1990. "Organizational Culture": American Psychologist, Vol. 45: February, pp. 109-119.

Senge, P. M., 1990. The Fifth Discipline: The Art and Practice of the Learning Organisation, Century Business: New York.

Sinkula, J. M., 1994. "Marketing Information Processing and Organizational Learning", Journal of Marketing, Vol. 58: January, pp. 35-45.

Slocum, J. W. Jr., McGill, M.E. & D. Lei, 1999. Designing Organisations for Competitive Advantage, Organisational Dynamics, Winter, pp. 24-38.

State, R., 1989. Organizational learning: The Key to Management Innovation. Sloan Management Review, Vol. 30, Spring: pp. 63-74.

Toffler, A., 1970. Future Shock: Bodley, London.

IFAC

Publications
www.elsevier.com/locate/ifac

COGNITIVE ANALYSIS AND SITUATION MODELLING

Maximov V.

*Maximov V. - Institute of Control Sciences of Russian Academy of Sciences,
Head of Sector of Cognitive Analysis and Situation Modelling, Ph.D.*

Abstract: Methodology, represented herein, synthesises system and cognitive approaches and represents the universal toolkit for structuring and understanding of complicated socio-economical systems behaviour. *Copyright © 2001 IFAC*

Keywords: Cognitive science, cognitive technologies, economic systems, forecasts, modelling

1. COGNITIVE STRUCTURING OF KNOWLEDGE ABOUT A SITUATION

At the heart of analytical technology of cognitive modelling lays cognitive structuring of knowledge about any object (region, state, etc.) and its external environment (economic, political, etc.), and object and external environment are differentiated "indistinctly".

The purpose of structuring is to reveal the most essential factors describing a "boundary" layer of interaction of object and external environment as a "world model", and establishment of qualitative relationships of cause and effect between them, i.e. what interference is exerted by one factor on an other during their change. Interference of the factors is displayed with the help of a cognitive map (model), which represents a sign (weighted) oriented graph.

Structuring of knowledge about regional economic situation problems includes creation of list of basic concepts, determination of the relations between them, setting of goals and determination of activities to be realised to achieve the goals. In other words, knowledge about a regional situation may be represented as a weighted oriented graph, table, text etc.

2. SPECIALITIES OF SITUATION CONTROL

The following problems are inevitably arise before the user when he analyses the present condition of any complicated socio-economic situation (http://www.ipu.ru/labs/lab51/51_home.htm):

1) What modifications of a situation are possible in (nearest) future?

2) What controls should be chosen for ensuring the desirable behaviour of the goal factors?

3) What problems thus can arise?

Problems of group 1 are connected with forecasting of strategy of possible modifications in a present situation. These modifications can be stipulated by internal reasons (for example, realisation of some control can be connected with the modification of interaction of the factors in a real situation and the similar modification can generate new problems) and by external reasons because a real situation is constantly exposed to exterior perturbations, which sources are not included in cognitive model of the situation to be analysed. The external reasons are divided into predictable, which origin can be foreseen using information from mass media and other sources, and on unpredictable. Irrespective of the character of changes in a situation, their presence requires modification of initial cognitive model of a situation.

Problems of group 2 are the problems of a routine (operative) economic situation control that is exercised in order to reach the goals. Solution of this problem can be performed by a few variants of "suitable" control. As was initially postulated, each concept of cognitive model, uniquely corresponded to situation specific concept, and the realisation of each variant of control provides realisation of the appropriate specific activities. Here arises a necessity of a comparative estimation of these variants by

- proximity of the results of control to the marked goal (on effectiveness of variants);

- expenditures (financial, physical, moral etc.) to be corresponded to realisation of each variant;

- character of consequences (converted, irreversible) after realisation of the appropriate variants in a real situation, etc.

Problems of group 3 are connected with the cognitive model analysis of the situation and exposition of problems to be generated in it (in particular, possible occurrence of crisis situations). New problems may be connected with ensuring of desirable behaviour of the changed goal factors in the changed situation. Thus analysis and solution of problems corresponding with a possibility of crisis situations should be made before real rise of such situations. That allows the user to undertake anticipatory activities to prevent the crisis situation, or to be "better" prepared for their overcoming.

3. CONCEPT AND COGNITIVE MAP OF REGIONAL SOCIO-ECONOMIC SITUATION

The cognitive map of a situation represents the oriented weighed graph, in which

- nodes are correspond to the basic factors of a situation, in terms of which the processes in a situation are described. The set of originally defined basic factors can be verified with the help of a data mining process, permitting to reject the "surplus" factors "poorly connected" to "kernel" of the basic factors;

- the direct correlation between the factors is determined by reviewing the cause-effect line-ups circumscribing distribution of influences from each factor to the others. It is considered, that the factors included in premise "if..." of the line-up "if..., then...", influence the factors of a corollary "that..." of this line-up, and this influence can be either strengthening (positive), or breaking (negative), or variable sign depending on possible side conditions.

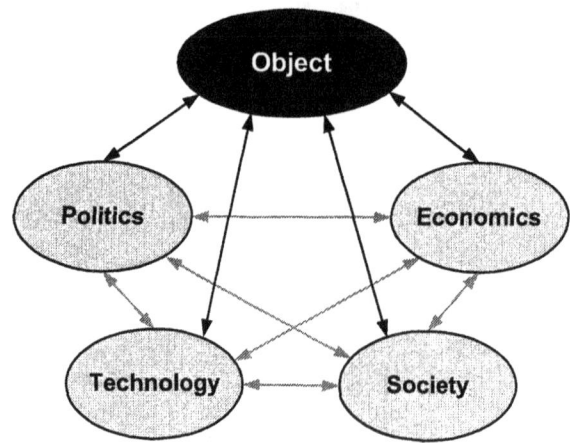

Fig. 1. PEST-analysis

Selection of the basic factors is carried out on the basis of PEST-analysis (Politics, Economy, Society, Technology), by means of which the political, economic, social, cultural, and technological aspects of external environment around of the researched object are analysed (Fig. 1). PEST-analysis is a tool of the usual four-element strategic analysis of external environment. Thus for each specific complex object there is a special set of the key factors, which directly and most significantly influences it. The analysis of each of the allocated aspects is carried out systematically, as far as all these aspects are closely and complexly interconnected. The significant change of any of aspects, as a rule, influences all chain. Such changes in each concrete case can create a threat to the object development, or, on the contrary, a new strategic opportunity of its future successful development.

The cognitive map depicts only the fact of influences of the factors on each other. It does not reflect a detailed character of these influences, dynamics of their modification depending on a modification of a situation and temporal modifications of factors. The account of all these circumstances requires passage to the following level of structuring of the information displayed in a cognitive map, i.e. to the cognitive model. At this level each connection between the factors of a cognitive map is uncovered up to the appropriate equation, which can contains both quantitative (measurable) and qualitative (not measurable) variables. The quantitative variables enter by a natural way as their numerical values. Each qualitative variable puts in correspondence a population of linguistic variables, mapping a various states of this qualitative variable (for example, the consumer demand can be "weak", "moderate", "agiotage", etc.), and the defined numerical equivalent in a scale [0,1] corresponds to each linguistic variable.

The following step is the situation problems analysis or SWOT-analysis (Strengths, Weaknesses, Opportunities, Threats). It consists of the analysis of strong and weak parties of researched object in their interaction with threats and opportunities of external environment (Fig. 2). Using SWOT-analysis the re-

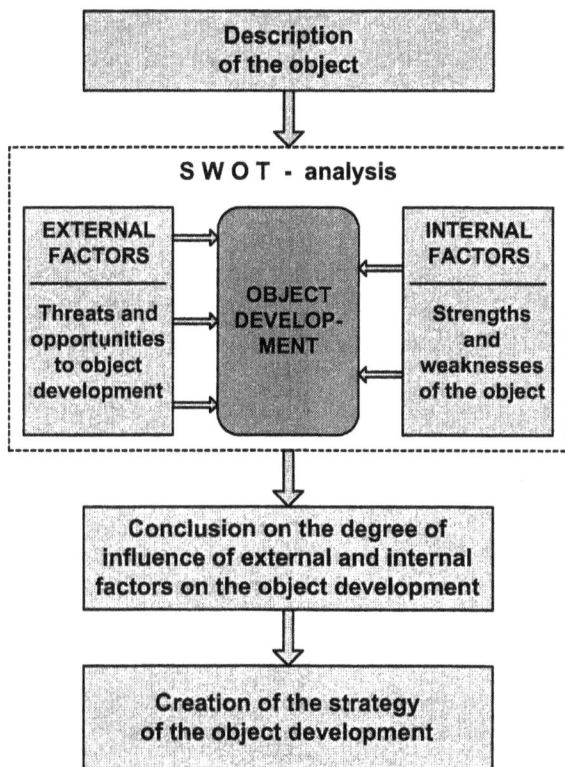

Fig. 2. SWOT-analysis

searcher obtains the possibility to determine urgent problem areas, bottlenecks, chances, and dangers to object under research. Opportunities are defined as something promoting to favourable development of the object. Threats is that can put damage to object, deprive its existing advantages. On the basis of the analysis of various combinations of strengths with threats and opportunities, and also weak parties with threats and opportunities, the problem field of researched object is formed. Problem field is a set of problems existing in the researched object in their interrelation with each other and with the factors of external and internal environment.

Results of PEST- and SWOT-analysis are used for model construction and modelling of control of the object development taking into consideration the model's cognitive "horizons".

The knowledge acquisition process can be divided into more "thin" processes (extraction, acquisition, shaping), having their own specificity. During extraction of knowledge there is an interaction of the expert (source of knowledge) and the cognitologist. It allows to trace a course of reasoning of the experts at decision-making and to reveal structure of their representations about a problem domain. As a result of their co-operation the cognitologist creates a "frame" model of a problem domain using his experience in cognitive psychology, system analysis, mathematical logic, etc.

Cognitive structuring, as far as it promotes the best understanding of problems, detection of inconsistencies and qualitative analysis of the economic, social

or political system, seems to be a convenient tool for research of macroeconomic environment, processes in the financial and credit market, supply and demand, behaviour of the competitors, etc. Practical achievements of the last years in the field of intelligent technologies have created a favourable basis so that the cognitive paradigm became attractive and popular. Now it promptly wins wide gratitude among authorities of states and regions, experts in business and economic management, and etc.

During the process of accumulation of knowledge about a situation under research, it becomes possible to uncover character of connections between the factors in more details. Here essential help can be rendered by use of data mining procedures.

Formally, the cognitive model of a situation may, as well as the cognitive map, be submitted by a graph, however each arc in this graph represents already certain functional association between the appropriate basic factors, i.e. the cognitive model of a situation is represented by a functional graph.

4. PROBLEMS OF PROVIDING OF GOAL BEHAVIOUR IN A COMPLICATED SITUATION

At the analysis of a concrete situation the user usually knows or assumes, what modifications of the basic factors are desirable for him. The factors representing the greatest interest for the user, shall be named goal ones. They are the "output" factors of cognitive model. The solution of control problem in a situation is to provide desirable tendencies of the goal factors, it is the core of control problem. The goal is considered to be correctly preassigned, if the desirable tendency of one goal factor does not cause undesirable tendencies of other goal factors.

In an initial set of the basic factors the population of the so-called controlling factors ("input" factors of the cognitive model) is selected. Control actions in the model are realized via these factors. The control action is considered to be co-ordinated with the goal, if it does not cause undesirable tendency in any of the goal factors.

If the goal of control is preassigned correctly and the control actions are co-ordinated with this goal, the solution of a control problem does not cause any specific difficulties (even in non-linear cognitive model of a situation with constant signs of influences of the factors). In a common case the determination of conditions of goal behaviour in a situation is a standalone problem and requires a special reviewing (Maximov V., Kornoushenko E., 2001).

5. MODELLING

Modelling is a mean of detection of economic, political and social regularities, warning and prevention of negative tendencies, deriving of theoretical and practical knowledge about a problem and formulation of practical conclusions on this basis.

Modelling is a cyclic process. Knowledge of a researched problem is extending and specifying, and the initial model is being constantly improved.

Modelling is based on the scenario approach.

Scenario consists of collection of factor tendencies describing a situation at the present moment, desirable goals, a set of activities that are used upon the running of a situation, and system of observable parameters (factors) illustrating behaviour of processes.

Scenario can be simulated in three basic directions:

1. Forecast of a situation self-development (without any actions to modify processes in a situation, when a situation is allowed to run its natural course);

2. Forecast of a situation development with the chosen set of activities (controls) (direct problem);

3. Synthesis of activities set to lead the situation to the desirable state (inverse problem).

Stages of modelling:

1. Definition of the initial conditions - tendencies describing development of a situation before modelling. This is necessary to make the model scenario more adequate to real situation;

2. Definition of goal in terms of desirable directions (increase, decrease) and strength (weak, strong) of goal factors tendencies;

3. Choice of a set of activities (collection of controlling factors) and definition of their possible and desirable strength and directedness of action on a situation;

4. Choice of the observable factors (indicators) describing development of a situation is carried out.

Three directions of modelling are represented in Tab.1.

Table 1

Stages of modelling	Directions of modelling		
	1. Situation self-development (development of a situation without any action on processes in it)	2 Development of a situation under chosen set of activities (controls) (direct problem solving)	3. Synthesis of a set of activities to reach the desirable state of a situation (inverse problem solving)
1. Definition of the initial state of a situation before modelling	+	+	+
2. Definition of goal factors, directions and strength of their tendencies	+	+	+
3. Choice of a collection of the controlling factors and strength of their action		+	
4. Choice of a set of possible actions, the force and directedness of which have to be determined			+
5. Choice of the observable factors (indicators)	+	+	+

6. DIALOG SOFTWARE PACKAGE "SITUATION"

The cognitive modelling technology may be realised with the use of dialog software package (DSP) "Situation" which has been developed for a structuring, qualitative analysis and obtaining of administrative solutions in complicated situations (economic, socio-political, regional, market, ecological etc.), where is the lack of the complete quantitative or statistical information.

DSP "Situation" allows to describe and justify the usual situation and to offer ways of reaching the goals with consideration of peculiarities of a specific situation on a qualitative level.

DSP "Situation" ensures:

1. Construction of cognitive model of a situation :

- Selection and substantiation of the basic factors of a situation;

- Establishment and substantiation of correlation of the factors;

- Construction of graph model of a situation.

2. Structural interpretation of problems requiring solution in the situation.

3. Searching and substantiation of strategy of goal reaching in stable or changing situations:

- choice and substantiation of the desirable goals in conditions of uncertainty;

- choice of activities (controls) necessary for reaching the goals;

- analysis of basic possibility of reaching of goals from an initial state of a situation with the use of chosen activities;

- analysis of restrictions on a possibility of realisation of the chosen activities in reality;

- analysis and substantiation of a real possibility of goal reaching;

- development and comparison of strategy of goal reaching.

4. Substantiation of possible scenarios of the situation development.

5. Machine generation of the reports and systematisation of results of a problem modelling.

REFERENCES

Maximov V., Kornoushenko E. (2001). Mathematical Basics of Construction the Graph and Computer Models for Complicated Situations. Preprints of the IFAC Symposium on Modelling and Control of Economic Systems SME 2001. (to be published).
http://www.ipu.ru/labs/lab51/51_home.htm.

IFAC
Publications
www.elsevier.com/locate/ifac

ASPECTS OF DECISION MAKING

Robert Genser

IFAC-Beirat Austria
Malborghetgasse 27-29,6/6, 1100 Vienna, Austria
Tel. & Fax: +43 1 6074187
rgenser@aon.at

Abstract: Attributes of efficient decision-making processes for complex systems considering human beings are outlined. The decision-making process is taken in account as a prerequisite that suitable measures are started for the solution of problems. The impact of new technology for knowledge management and learning is stressed. Object oriented modelling and visual presentations are pointed out. *Copyright © 2001 IFAC*

Keywords: Decision making, committee, complex systems, knowledge representation, learning, object-oriented, formal methods, visual.

1. INTRODUCTION

The decision-making process for solving problems in complex systems has to consider the behaviour and capacity of human beings. The way of decision making influences very strong if actions will be taken to solve the problem.

The process for arms control and arms reduction as well as the striving for a sustainable environment for example are analysed in depth already, see (Kremenyuk, 1991) and (Victor, 1995).

Supporting tools are developed for improving decisions of committees dealing with complex problems (Wierzbicki et. al, 1999). But many approaches are sticking to mathematical models not taking in account the demand of reality. The progress of technology stimulates new ways of knowledge

management (Tochtermann and Maurer, 2001) for improving human actions.

2. HUMAN ASPECTS IN DECISION MAKING

The objectives are the basis for decision making. They may be conscious in case of rational decisions. But many paradigms neglect the change of aims by situation, see Figure 1, or by satisfaction of some objectives, like appeasing human hunger. In many cases, a director of a company has to decide for the design of a product even he would be against this decision as individual.

It is assumed that decisions and actions of human individuals are rational. But the problem of alcohol drinking of car drivers gives the counter-evidence.

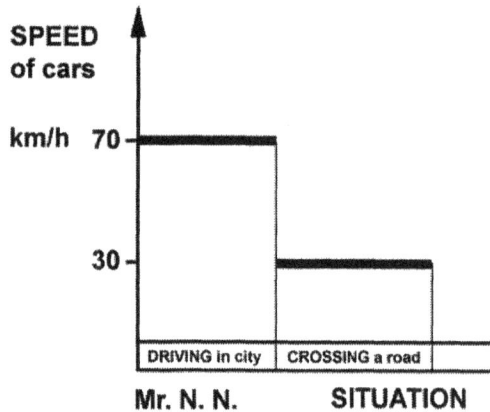

Fig. 1. Change of objective of a person depending on their situation in traffic

Not lack of knowledge rather lack of consciousness causes bad results. But as hearing does not mean listening, listening not understanding, understanding not agreement, agreement not commitment, commitment does not warrant sustainability.

Learning is a way of improving knowledge. But knowledge is not a collection of objective facts. Learning textbooks by heart does not gain rational actions. A sequence of events and actions are needed for improving consciousness and skills. Population with experience of real bad sides of war is more prone to find solutions for peace.

Many barriers are recognised which cause irrational behaviour of individuals and organisations. Literature on management and innovation processes is dealing more or less exhaustively with these human weaknesses. Some of these topics are given for example:
- prejudices
- lack of confidence
- bad experience
- lack of sensitivity with dealing with others
- lack of communication skill
- too much concern for present opinion of followers or superior
- concern that also increase of advantages for others or jealousy
- lack of time or resources
- to take all to seriously.

The results of a study by members of the Stanford University on the influence of cultural environment on human behaviour can be found in (Cialdini, 2001). It depends on the culture, if an individual responses according own experience, friends, loyalty to high-ranking persons, or regulations given.

Conflicts are stimulating irrational behaviour. But conflicts are inevitable in real life and are not substantial a thread for a sustainable development. Conflicts should be approached in an open manner and not be covered up. They should be solved in an early stage and not be avoided. But a solution should not be enforced. This is not a new experience, but actors do not obey this truism, example given: multi-cultural problem.

As Prof. Tom B. Sheridan stated many years ago concerning acceptance of automation by society, people need trust for rational behaviour. Trust needs clear information and not diplomatic wording.

Besides the lot of literature on management at business level, also experience at diplomatic level is published, see for example (Kaufmann, 1988).

The problem of misunderstanding or not identical interpretation of meaning of words is quite frequently. In standardisation bodies it is an initial process to get agreement on terminology before real work can start. But it is a learning process for the group because some fuzziness of the description of the real world has to be recognised. Also in documentation of software it was learned that a full description in written form is not possible without depending on context of experience in brain.

The aspiration of members and important information is not transparent or even not recognisable in committees and rational learning processes are suppressed. Human limitations, which can not be improved further by organisational or social framework, may be very well reduced by technology as it is shown in many areas of daily life.

3. TECHNOLOGICAL ASPECTS

It is agreed that trust and acceptance of groups involved can be gained by **participation** in decision making process with **learning feedback** (Genser, 1986). The International Institute for Applied System Analysis in Laxenburg started already about 1983 to improve the problem of large selecting committees with developing a prototype for complex transportation planning for the Austrian Federal Ministry for Public Economy and Transport. To improve the information transfer it was tried to use visual presentations and interactivity instead of evaluating single numbers and using complex inflexible models. As a starter for a learning process between the members the **aspiration** levels are made transparent, see Figure 2, and analysed (Granat and Makowski, 2000).

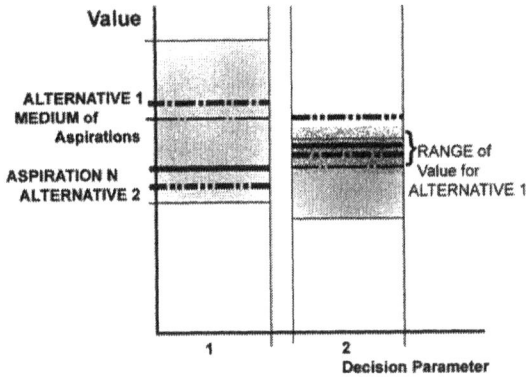

Fig. 2. Aspiration levels and value of alternatives

Also different levels of knowledge or different opinions should be become known by this transparency and can initiate further investigations. Simulation, pilot projects etc. may be used for getting missing knowledge.

Sophisticated mathematical algorithms are not suitable for fuzzy information available and according the complexity of problems (Genser, 2000). **Object oriented formal methods** have to be applied (Jones, 1990; Rumbaugh et al., 1991). Simulation models should use **metaphors** instead of hard data (Hopkins et al., 2001).

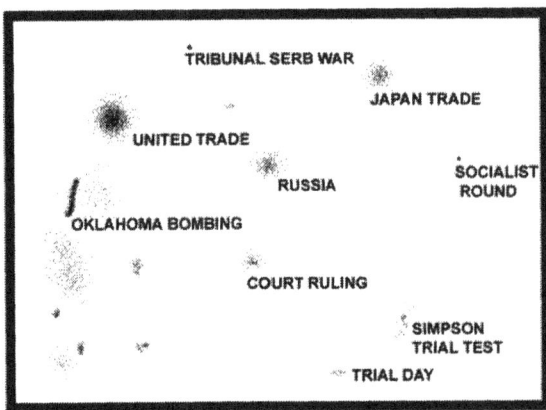

Fig. 3. Visual presentation of knowledge

An interesting approach for discovering knowledge by visual analysis is shown by Thomas et al. (2001). The embedding of terms and their relations in different documents is analysed and these connections are presented like star clusters on the sky with different colours, see Figure 3. The connections can be recognised with one glimpse and the documents can be accessed directly.

Much effort can be recognised not only to develop suitable data presentation at the human-machine interface, but also to implement **emotion** (Miyatake, et al., 1998) in information transfer to human beings. Much can be learned by the experience of the movie industry in Hollywood.

The embedding of a decision-making support module is shown in Figure 4. The system for getting decisions and for improving acceptance as well as actions should stimulate learning between the committee members. It should be flexible concerning access to knowledge according the particular demand. The existing information should be transparent and presented according the requirements of human beings. Efficient databases and ad hoc coarse simulation models are better than huge sophisticated models. The decision process itself should be improvable by learning feedbacks.

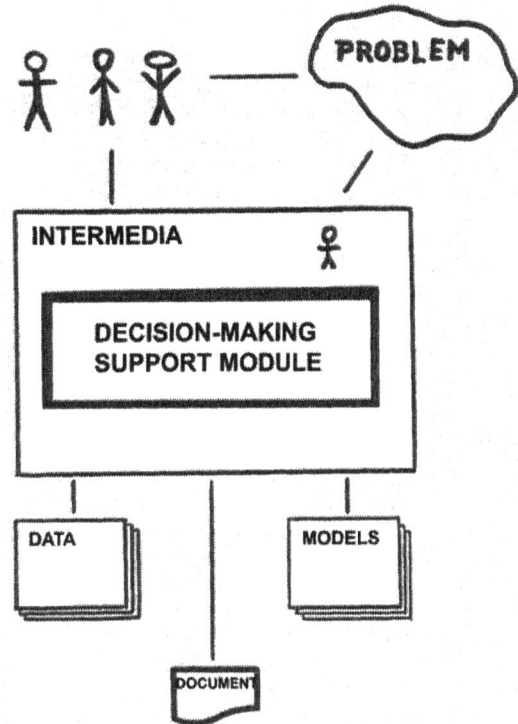

Fig. 4. Embedding of a decision-making support Module

SHINAYAKANA (Sawaragi, et al. 1990; Nakamori, et al., 2000) is an approach in Japan for making decisions in complex areas. This system comprises administrators, citizens, enterprises, international organisations, media and researchers. Hard data are gathered at the scientific level and a framework model as well as soft modelling techniques are developed. Human attitude and behaviour are investigated also using the Web. A multi-agent simulation platform is constructed.

Reasonable **standardisation** is needed for efficient use of technology. For example, a standard for learning object metadata is developed by IEEE (Duval, 2001). Standardised methods are a prerequisite for having enough skill and reduced effort for learning.

4. CONCLUSION

Guidance by social or governmental framework is used for improving human behaviour. But further enhancement is possible by applying technology already available.

A good decision process results actions to solve problems rather to decide for doing nothing.

REFERENCES

Cialdini, R. B. (2001). Influence across Cultures. *Scientific American,* **284**, 67.

Duval, E. (2001). Metadata Standards: What, Who & Why. In: *Proceedings of I-KNOW`01* (Tochtermann, K. and H. Maurer, Eds.), 137-147, Springer-Verlag, Wien.

Genser, R. (1986). Learning in Decision Making. In: *Large-Scale Modelling and Interactive Decision Analysis* (Fandel, G., Ed.), 138-147. Springer-Verlag, Berlin.

Genser, R. (2000). Dealing with fuzzy information in real world. In: *Preprints of the IFAC Workshop on Supplemental Ways for Improving International Stability*, 30-34, Ohrid.

Granat, J. and M. Makowski (2000). *Interactive Specification and Analysis of Aspiration-Based Preferences.*RR-00-09, IIASA, Laxenburg.

Hopkins, J. F. and P. A. Fishwick (2001). A Three-Dimensional Human Agent Metaphor for Modeling and Simulation. *Proceedings of the IEEE,* **89**, 131-147.

Jones, C. B. (1990). *Systematic Software Development using VDM.* Prentice-Hall, Englewood Cliffs, NJ.

Kremenyuk, V. A. (Ed.) /1991). *International Negotiation - Analysis, Approaches, Issues.* MacMillian International, Bristol.

Miyatake, T. and K. Shimohara (1998). Autonomous graphics generation system for supporting human self-expression. *Preprints of the 7th IFAC/IFIP/IFORS/IEA Symposium on Man-Machine Systems*, 473-478, Kyoto.

Nakamori, Y. and Y. Sawaragi (2000). Methodology of Knowledge Integration and Creation for Environmental Issues. *14th JISR-IIASA Workshop on Methodologies and Tools for Complex System Modeling and Integrated Policy Assessment,* Laxenburg.

Rumbaugh, J., M. Blaha, W. Premerlani, F. Eddy and W. Lorensen (1991). *Object Oriented Modelling and Decision.* Prentice-Hall, Englewood Cliffs, NJ.

Sawaragi, Y., M. Naito and Y. Nakamori (1990). Shinayakana systems approach in environmental management. *Preprints of the 11th IFAC World Congress*, 282-287, Tallinn.

Thomas, J., P. Cowley, O. Kuchar, L. Nowell, J. Thomson and P. Ch. Wong (2001). Discovering Knowledge Through Visual Analysis. In: *Proceedings of I-KNOW'01* (Tochtermann, K. and H. Maurer, Eds.), 62-74, Springer-Verlag, Wien.

Tochtermann, K. and H. Maurer (Eds.) (2001). *Proceedings of I-KNOW'01*, Springer-Verlag, Wien.

Victor, D. G. (1995). *Design Options for Article 13 of the Framework Convention on Climate Changes: Lessons from the GATT Dispute Panel System.*ER-95-1, IIASA, Laxenburg.

Wierzbicki, A., M. Makowski and J. Wessels (Eds.) (1999). *Model-Based Decision Support Methodology with Environmental Applications*, Kluwer Academic Publishers, Dordrecht.

IFAC

Publications
www.elsevier.com/locate/ifac

TECHNOLOGICAL CHANGE AND PERFORMANCE OF OPTIMAL ECONOMIC POLICY: SOME RESULTS FOR AUSTRIA

Gottfried Haber*

* University of Klagenfurt, Austria

Abstract: Inside the European Monetary Union (EMU), there are efforts to reduce public debt to fulfill the criteria imposed by the Stability and Growth Pact (SGP), which are mandatory for members of the Euro zone. Due to these constraints, public expenditure has been cut in several European countries. Thus, the question of the effects on technological change arises. In this paper, the economic implications for optimal budgetary stabilization under alternative rates of technological change are evaluated for Austria, in order to assess the importance of technological progress for stabilization policy. Copyright ©2001 IFAC

Keywords: Optimum Control, Economic Policy, European Monetary Union, Fiscal Policy, Technological Change

1. INTRODUCTION

Technological change is one of the main sources of increasing welfare and a source of social stability, as it leads to a (theoretical) possibility of redistributing economic output without the need for reducing wealth of any economic subject in the economy ("Pareto" improvement). On the other hand, technological progress influences both the effects and the sustainability of economic policy as it changes reaction patterns of economic aggregates. Three main questions for policy makers arise: (a) the economic effects of technological progress on a macro level, (b) the microeconomic effects (allocation and distribution), and (c) implications for optimal economic policy. This paper uses an econometric model of the Austrian economy (AMOD1) and the optimization algorithm OPTCON to treat the third question for Austria.

As in several other European countries, public debt in Austria had been rising during the last three decades until the introduction of the common currency zone in Europe. The peak level of public debt was nearly 70 percent of GDP at the end of the twentieth century, which is significantly above the 60 percent reference value given by the SGP. Only relatively high priority on the debt target might lead to a sustained consolidation of the public sector (Haber, 2001); these results are used to restrict the number of optimization excercises to a so called strong consolidation scenario. Exogenous productivity changes caused by technological progress are imposed on the model. Optimal economic policy is calculated for those different scenarios. In particular, optimal stabilization of public debt is investigated. As previous papers have shown, fiscal consolidation efforts tend to produce significant output losses (welfare losses) in the short-run, but in the long-run even positive effects can be observed. Some results for European monetary policy have been described previously (Neck et al., 2000); this paper concentrates on the fiscal policy aspect.

2. FRAMEWORK: AMOD AND OPTCON

AMOD1 (Haber, 2000) is an annual econometric model of the Austrian economy, which is well suited for economic policy analyses. The version of the model used in this paper consists of 57 equations, including 29 (stochastic) behavioral equations and 28 definitions. The number of estimated coefficients is 97; in the base version of AMOD1, 13 exogenous variables are included, where four of them may be used as economic policy instruments. AMOD1 is based upon neoclassical assumptions as well as on Keynesian considerations: There are (partly) forward-looking expectations on the one hand and backward-looking rigidities on the other hand. As a result, the model provides a well-defined long-run growth path and disequilibria at the same time. The main sectors are: a twice disaggregated consumption block, the wage-price system, a production sector, the monetary sphere, a foreign trade sector and, the public sector. For all parts of the model, microeconomic foundations and optimizing behavior of the private agents have been applied to derive the specifications of the estimated equations. The highly nonlinear model is estimated in blocks, using iterative three-stage least squares (iterative 3SLS), seemingly unrelated regressions (SUR) and ordinary least squares (OLS). Estimation period is 1976–1998; extensive evaluation and checks of all parameters have been performed (Haber, 2000) and show that the quality of AMOD1 may be compared to other widely used models of the Austrian economy.

Private consumption is determined in a consumption block that is disaggregated twice, in eleven groups of goods (food and beverage, health, housing, etc.) and four types of goods (durables, non-durables, etc.). The specification is compatible with the absolute income hypothesis and the permanent income hypothesis, allowing for Keynesian as well as neoclassical elements. The wage-price system is highly interdependent and comprises the main factors of inflation as well as the specific institutional situation of the Austrian labor market and the characteristics of the wage bargaining process. Inflation may be effected by wage increases, by higher import prices and by high capacity utilization. Nominal wages in turn are determined by the price level, by the level of unemployment and by changes in productivity. Economic agents exhibit partly forward-looking behavior in both determining prices and wages. Labor supply is disaggregated by gender. As a factor of production, labor is treated as a homogenous input. Labor demand is derived from the first order conditions of profit maximization of the firm. The specification of private investment is based upon the concept of the maximization of the market value of the firm, which is compatible with Tobin's marginal q and with the extended accelerator hypothesis.

Potential output Y, as derived from equation 1 (K is the capital stock, γ is the natural growth

rate of potential GDP and includes technological progress, A_0 is the constant of the Cobb-Douglas production function, t is a time trend, L is employment, and $(1 - \alpha)$ is the wage quota), is calculated using an estimated constant elasticity of scale (CES) production function of the Cobb-Douglas type by substituting actual labor demand with the total labor force. For the optimizations here, the coefficient γ is modified at the beginning of the optimization period. As the base year for the linear trend t is 1970 and the beginning of the simulations in this paper is 2000, the γt is replaced by equation 2, where $\beta = \gamma$ and δ is a productivity offset:

$$\ln(\frac{Y}{K}) = A_0 + \gamma t + (1 - \alpha)\ln(\frac{L}{K}), \qquad (1)$$
$$\gamma t = \beta t + \delta(t - 30), \qquad (2)$$

Inversion of the real money demand equation yields a nominal short-run interest rate equation, which is estimated directly. The exchange rate peg to Germany (since the beginning of the eighties) enters the model through the historical values of the (effective) exchange rate and money supply. Based upon the short-run interest rate, both a long-run interest rate and an effective interest rate for the public sector are calculated. This accounts for the realistic feature that the interest rate on public debt is different from the interest rates for the private agents, at least in the short-run, and is especially useful for analyses of the public sector.

Exports are mainly determined by the relative competitiveness of domestic production, while imports essentially depend on domestic demand. For the chosen specification, income effects (in the export equation) and price effects (in the import equation) could be identified as rather small and were therefore dropped. On the foreign goods markets, rigidities are explicitly taken into account. The (nominal effective) exchange rate equation is built around the weak interest rate parity hypothesis (for the determination of the exchange rate), which is a reasonable paradigm in the presence of high capital mobility. For the estimation period, it is assumed, that Austria faces flexible exchange rates to the rest of the world (except Germany and some other countries implementing an exchange rate peg to Germany).

Optimum control of economic systems can be used for evaluating economic policy and to find "optimal" values for the economic instruments. This makes sense, if there are given desired target values for at least some of the endogenous variables (an "ideal situation". Weights have to be attached to the objectives, indicating their relative importance. Thus, an objective function can be specified, for ease of computation generally in the form of an additively separable quadratic welfare loss function, which is minimized subject to the econometric system.

OPTCON (Matulka and Neck, 1992) is an optimization algorithm that is able to deal with both deterministic and stochastic setups (with uncertainty in the parameters and in the residuals) for linear and non-linear problems, which distinguishes OPTCON from other widely used algorithms (Chow, 1975; Chow, 1981; Kendrick, 1981). Due to limitations in previous versions of OPTCON concerning highly non-linear systems of medium to large scale, version 3.0001b (Haber, 2000) is used for the (deterministic) optimizations in this paper. In the original version, OPTCON uses a Gauss-Seidel procedure to approximate a deterministic solution of the system. Then, the non-linear system is optimized. This is done starting with the previously calculated approximation by iteratively running a backward recursion and applying Bellman's principle of optimality in order to get the parameters of a policy feedback rule. In a last stage, a forward projection (simulation) of the model is performed. In the version of OPTCON used in this paper, the Gauss-Seidel procedure is replaced by a modified Newton-Raphson algorithm, which tends to be more stable in the presence of extensive non-linearities. For further information refer to the sources mentioned above.

3. OPTIMIZATION SCENARIOS

For the experiments in this paper, first a baseline solution of the model is calculated, where the natural growth rate of GDP is kept at the original value of 0.0097 (this is approximately 1 percent). Thus, the offset δ is zero for the baseline. This simulation gives the desired values for the target variables in the subsequent optimization exercises. Table 1 shows the weights in the linear quadratic welfare loss function, which is imposed in the optimization runs.

Table 1. Weights of the targets in the objective function.

Variable	Dimension	Weight
Public Debt [ratio to GDP]	percent	1.000,000.0
GDP (real)	Mio. ATS	1.0
Private consumption (real)	Mio. ATS	1.0
Inflation rate	percent	100,000.0
Public consumption (nom.)	Mio. ATS	0.1
Public investment (nom.)	Mio. ATS	0.1
Net taxes, exogenous (nom.)	Mio. ATS	0.1

The main target is the debt to GDP ratio with a desired value of 60 percent. GDP and private consumption are minor targets of the economic policy maker in the scenarios considered here. Note that the instruments (nominal public consumption, nominal public investment, and the nominal

exogenous component of net taxes) are included in the objective function in order to increase the stability of the solution and the smoothness of the implemented optimal policy. The huge differences in the dimensions of the respective weights are given by the relative importance of the respective targets and by differences in the dimensions used for the variables. Apart from the debt to GDP ratio, all target values are taken from the baseline.

With this setup, three different optimization scenarios are run. Scenario 1 corresponds to the strong consolidation scenario mentioned above (Haber, 2001), without any modifications. Scenario 2 depicts an increase in technological progress leading to an increase in the natural GDP growth rate by 0.25 percentage points per annum (fast scenario). A reduction of the annual GDP growth rate by 0.25 percentage points is the framework for scenario 3 (slow scenario). The time horizon for all calculations is 2000–2010; in order to ensure model convergence and convergence of the optimum control solutions, more periods (up to the year 2025) were used for some of the calculations, where appropriate.

4. RESULTS

The main results of the fiscal consolidation in scenario 1 (reference optimization) are cuts in public expenditure (public consumption and public investment) and increases in net taxes. These measures decrease public debt gradually to the desired level of 60 percent of GDP; fluctuations of the deficit figure are still observable in the optimal time path. This very restrictive strategy of the policy maker leads to losses in output (GDP), lower inflation and a small increase in unemployment (Haber, 2001). Here, the point of interest is the sensitivity of the optimal fiscal consolidation policy with respect to the magnitude of technological progress. Note that most of the figures depicted in the following section represent deviations from the simulated baseline values. The horizontal axis is the time axis. In the legend of the figures, the digits 1–3 represent the calculated scenarios.

Figure 1 shows the time paths of the public debt to GDP ratio. In all three cases, the target value is reduced continuously (compared to the baseline, which is not given here), which indicates that the main target is followed more or less successfully for all degrees of technological change. Up to the year 2007, faster technological progress implies that the consolidation process can be done more easily. After this period, only the slow scenario (3) is able to reduce the stock of public debt even below the reference value, which of course also has to be seen as suboptimal from the point of

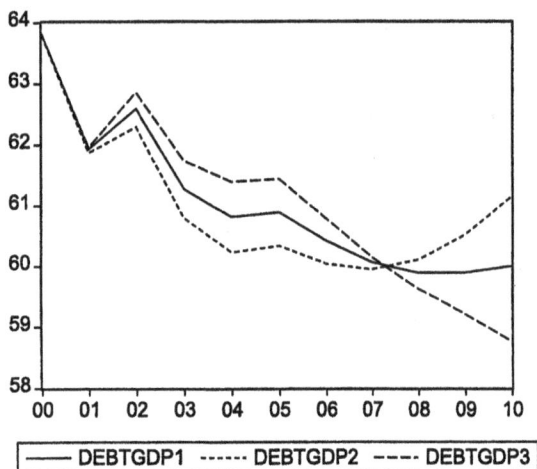

Fig. 1. Public debt ratio (level) [percent of GDP]

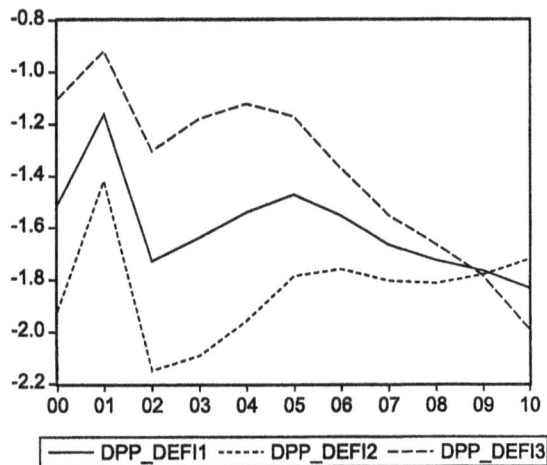

Fig. 2. Deficit ratio (deviation) [percentage points]

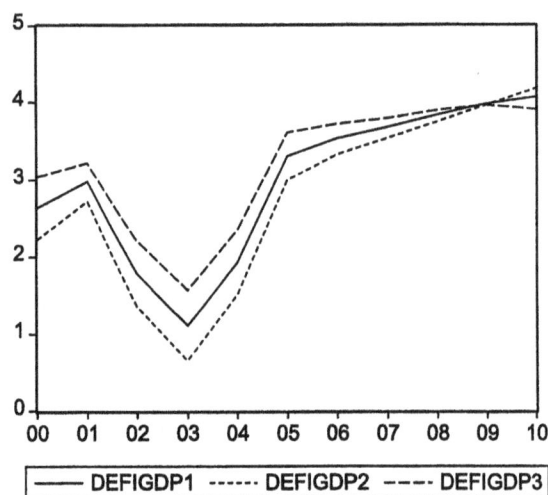

Fig. 3. Deficit ratio (level) [percent of GDP]

the objective function. This result seems counterintuitive at this stage but cannot be interpreted without looking at the optimal values of the economic policy instruments. As will be shown below, faster technological progress implies that the economic policy measures implemented by the policy maker may be less expansionary or, as in this case, more contractionary. Thus, (nearly) successful fiscal consolidation might be reached earlier, but on the other hand this more intensive consolidation might reduce the stock of capital in the long run compared to the reference optimization (1) and the slow scenario (3), leading to lower GDP. As GDP is the denominator of the debt to GDP ratio target value, this arithmetic might be the solution to this phenomenon, which is of course related to the discussion of Keynesian and non-Keynesian effects of fiscal policy. Additionally, and that might be the main effect, the reductions in inflation and the price level (see below) lead to an increase in debt in real terms. As real GDP cannot be increased beyond potential output in the long run, stabilization of the debt to GDP ratio might become more difficult with subsequent periods. These findings should be investigated in more deatail in future analyses.

Figures 2 and 3 show the development of fiscal deficits (the flow variable related to the stock of public debt). In all three scenarios, deficit fluctuations are preserved, leading to smoother results for the debt, private consumption, and GDP targets. Until the year 2009, deficits (as percentages of GDP) are lower for higher degrees of technological progress. In turn, an explanation to this reversal of the three scenarios at the end of the optimization horizon might be based upon the effects on the price level.

The optimal values of the policy instruments (nominal public consumption, nominal public investment and nominal exogenous net taxes) show that the optimal policy must be contractionary

in order to reach the fiscal target. Public expenditure is reduced significantly, while an increase in net taxes can be observed. As a rule of thumb, higher rates of technological progress lead to more restrictive measures. Again, this result has to be interpreted in the context of the target variable: higher growth rates tend to increase real output and reduce the debt ratio without the need for sharp restrictive policy measures. But on the other hand, higher potential output caused by technological progress increases the output gap (the difference between actual GDP and potential GDP) and tends to lower the price level. This leads to deflation (or lower inflation) and has negative effects on the business sector of the economy. Moreover, the lower price level increases real variables (e.g. public expenditure) ceteris paribus, thus the nominal magnitudes can be reduced without a change in real acitivity. Put differently, the price effects of higher rates of technological progress lead to more restrictive measures concerning the nominal instruments.

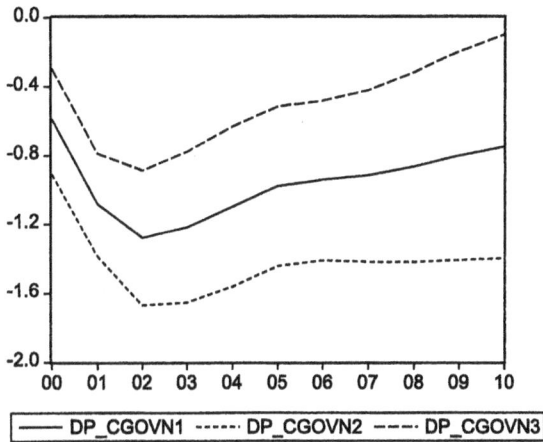

Fig. 4. Public consumption [percent of GDP]

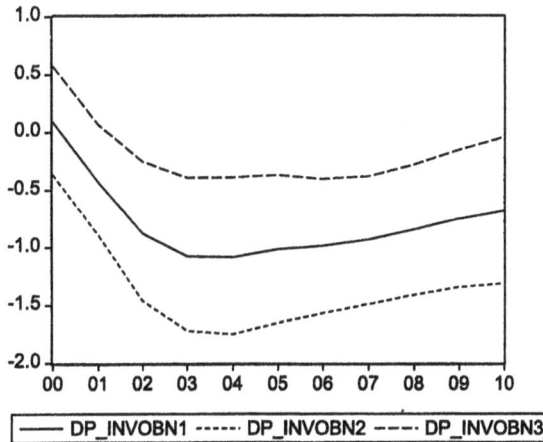

Fig. 5. Public Investment [percent of GDP]

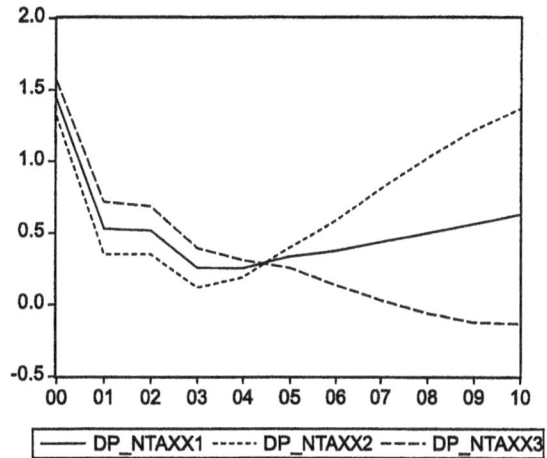

Fig. 6. Exogenous net taxes [percent of GDP]

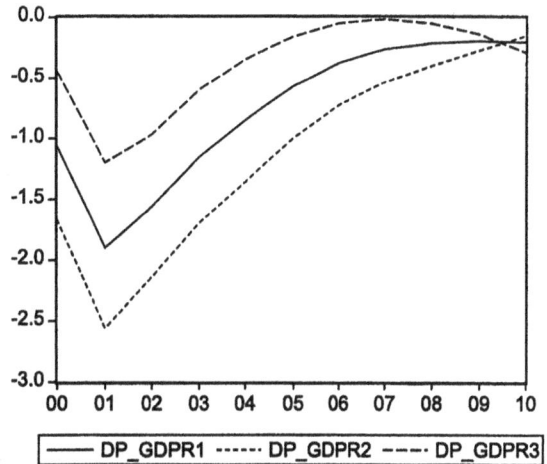

Fig. 7. GDP [percent]

Nominal public consumption (figure 4) is reduced by more than 0.8 percent of GDP in the slow scenario and by more than 1.6 percent of GDP in the fast scenario. In scenario 3 (slow), the instrument nearly returns to the baseline values, while the reduction in public consumption is permanent in the fast scenario and remains about 1.4 percent below the baseline in the long run.

Nominal public investment (as shown in figure 5) follows a similar time path as public consumption described above. The main difference is that in the first two periods, an increase in public investment can be observed for the slow scenario. This is consistent with theoretical considerations, as a slow-down of technological progress implies higher public activity in research and development. Public investment is lowest in the fast scenario and highest in the slow scenario.

In figure 6, the time paths for nominal exogenous net taxes are depicted. Taxes are raised by approximately 1.5 percent of GDP in the first period in all three scenarios and then gradually reduced towards the baseline. Taxes are generally slightly higher in the slow scenario until 2004; afterwards significant differences among the time paths can

be observed, where taxes are higher in the fast scenario now.

The main economic variable is real GDP (figure 7), describing the real output of the whole economy. GDP assumes rather different values in the three alternative scenarios: While output losses in the short run (maximum in 2001) amount to more than 2.5 percent in the fast scenario, they remain below 1.25 percent in scenario 3. In all three scenarios, output losses diminish gradually in the medium run, with the most preferable values in the slow scenario. At the end of the time horizon, the order of preference changes: Now scenario 2 produces the smallest output losses due to the fiscal consolidation program. Note that the relatively high output losses in 2001 are consistent with the mainstream forecasts for Austrian GDP growth for this period.

The high-tech scenario (2) shows inflation rates (figure 8) well below the baseline of the model (up to 1.4 percentage point lower in the short run and about 1 percentage point lower in the long run). In the reference optimization (scenario 1), the reduction of the inflation rate reaches a

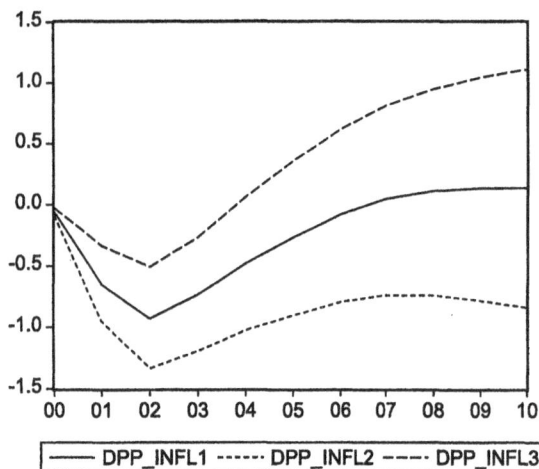

Fig. 8. Inflation rate [percentage points]

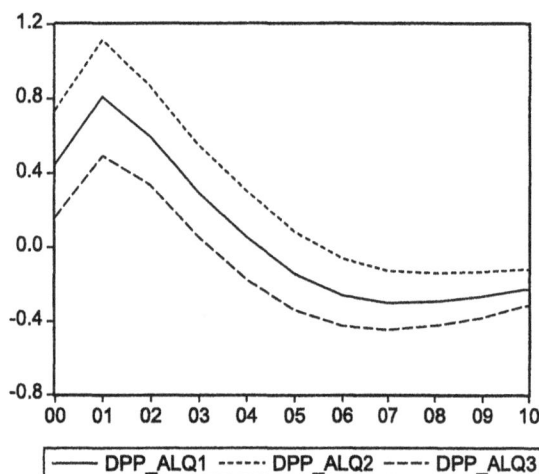

Fig. 9. Unemployment rate [percentage points]

maximum of −1 percentage point; in the long run, inflation returns to the baseline values. In the slow scenario (3), the inflation rate is slightly reduced during the first three periods but rises steadily and consolidates at about 1 perentage point above the baseline. Higher potential output increases the output gap and leads to deflationary effects in the short run. Although the inflation rate is includued in the objective function, stabilization of the inflation rate occurrs rather slowly, leading to cumulative differences in price levels among the baseline and the three scenarios analyzed here. This causes shifts in the nominal and real magnitudes of several key variables of the economic system and has already been identified as a source of some of the findings above.

In the fast scenario, unemployment (figure 9) rises by up to 1.1 percentage points above the baseline (year 2001). The peak in the slow scenario is only 0.5 percentage points. This can be attributed to the more restrictive policy measures in the scenario with a higher rate of technological progress, which tend to exhibit strong Keynesian effects in the short run. Note that after 2005, unemploy-

ment is below the baseline in all scenarios. This shows that (at least in the long run) employment returns to an equilibrium level, regardless of the specific policy measures taken. On the other hand, in the short run, employment reacts very sensitive to different rates of technological change in a fiscal consolidation framework.

5. CONCLUSION

For all periods, the reference optimization (scenario 1) produces values of the endogenous variables and of the instruments, which are in between the values of the other scenarios. Thus, the order of the scenarios is consistent for the whole time horizon. The analysis has shown that there are significant differences in the reaction of the economy and in the optimal time paths of the economic policy instruments, if the rate of technological change is altered in the model. Strong effects could be observed originating from changes in inflation and the overall price level, leading to shifts in the framework of nominal and real variables.

It is worth noting that some of the findings in this paper have not been treated in the economic literature, yet. Of course, the results presented here have to regarded as still preliminary and more extensive work on these issues should be performed. Especially, stronger theoretical foundations of the effects found in this paper and sensitivity analyses should be carried out. Nevertheless, this paper has shown the strong dependency of the specific design and the performance of optimal economic policy on the degree of technological change.

REFERENCES

Chow, G.C. (1975). *Analysis and Control of Dynamic Economic Systems*. New York et al.

Chow, G.C. (1981). *Econometric Analysis by Control Methods*. New York et al.

Haber, G. (2000). AMOD1. Ein makroökonometrisches Modell für Österreich. PhD thesis. Vienna University of Economics and Business Administration. Vienna.

Haber, G. (2001). Simulation analysis of public debt and fiscal deficit in Austria — optimal economic policies for the past and the future. Accepted for publication in Empirica.

Kendrick, D. (1981). *Stochastic Control for Economic Models*. New York et al.

Matulka, J. and R. Neck (1992). Optcon: An algorithm for the optimal control of nonlinear stochastic models. *Annals of Operations Research* **37**, 375–401.

Neck, R., G. Haber and W.J. McKibbin (2000). Macroeconomic policy design in the European Union: A numerical game approach. *Empirica* **26**(4), 319–335.

AUTHOR INDEX

www.ingramcontent.com/pod-product-compliance
Lightning Source LLC
Chambersburg PA
CBHW082307210326
41598CB00028B/4463